AF389008

Qualitative Analysis of Food Products

Qualitative Analysis of Food Products

Editor

Theodoros Varzakas

MDPI • Basel • Beijing • Wuhan • Barcelona • Belgrade • Manchester • Tokyo • Cluj • Tianjin

Editor
Theodoros Varzakas
Food Science and Technology
Theodoros Varzakas
Kalamata
Greece

Editorial Office
MDPI
St. Alban-Anlage 66
4052 Basel, Switzerland

This is a reprint of articles from the Special Issue published online in the open access journal *Foods* (ISSN 2304-8158) (available at: www.mdpi.com/journal/foods/special_issues/Food_Products).

For citation purposes, cite each article independently as indicated on the article page online and as indicated below:

LastName, A.A.; LastName, B.B.; LastName, C.C. Article Title. *Journal Name* **Year**, *Volume Number*, Page Range.

ISBN 978-3-0365-1846-6 (Hbk)
ISBN 978-3-0365-1845-9 (PDF)

Contents

About the Editor

Theodoros Varzakas

Theodoros Varzakas is a Senior Full Professor at the Department of Food Science and Technology, University of Peloponnese, Greece, specializing in issues of food technology, food processing/engineering, and food quality and safety. Section Editor in Chief Journal Foods in Food Security and Sustainability (2020-). Ex-Editor in Chief Current Research in Nutrition and Food Science (2015-2019). Reviewer and member of the editorial board in many international journals. Has written more than 200 research papers and reviews, and has presented more than 160 papers and posters at national and international conferences. He has written and edited six books in Greek, and six in English on sweeteners, biosensors, food engineering, and food processing, published by CRC. He has participated in many European and national research programs as a coordinator or scientific member.

Preface to "Qualitative Analysis of Food Products"

Qualitative control and analysis of food products is a requirement for food industries, both in terms of quality assurance and food safety management systems. Analysis of foods is continuously requiring the development of more robust, efficient, sensitive, and cost-effective analytical methodologies to guarantee the safety, quality, authenticity, and traceability of foods in compliance with legislation and consumers'demands.

Different analyses include microbiological and chemical analyses, from simple to complex, from old to modern technologies. Hence, fundamental and/or state-of-the-art methods of the development, optimization, and practical implementation in routine laboratories, and validation of these methods for the monitoring of food safety and quality, are employed. Methodologies for food microbial contaminants, food chemistry and toxicology, food quality, food authenticity, and food traceability have been presented and discussed in this Special Issue.

Theodoros Varzakas
Editor

Article

Polyphenolic Characterization and Antioxidant Activity of *Malus domestica* and *Prunus domestica* Cultivars from Costa Rica

Mirtha Navarro [1],*, Ileana Moreira [2], Elizabeth Arnaez [2], Silvia Quesada [3], Gabriela Azofeifa [3], Felipe Vargas [1], Diego Alvarado [4] and Pei Chen [5]

[1] Department of Chemistry, University of Costa Rica (UCR), Rodrigo Facio Campus, San Pedro Montes Oca, San Jose 2060, Costa Rica; luis.vargashuertas@ucr.ac.cr

[2] Department of Biology, Technological University of Costa Rica (TEC), Cartago 7050, Costa Rica; imoreira@itcr.ac.cr (I.M.); earnaez@itcr.ac.cr (E.A.)

[3] Department of Biochemistry, School of Medicine, University of Costa Rica (UCR), Rodrigo Facio Campus, San Pedro Montes Oca, San Jose 2060, Costa Rica; silvia.quesada@ucr.ac.cr (S.Q.); gabriela.azofeifacordero@ucr.ac.cr (G.A.)

[4] Department of Biology, University of Costa Rica (UCR), Rodrigo Facio Campus, San Pedro Montes Oca, San Jose 2060, Costa Rica; luis.alvaradocorella@ucr.ac.cr

[5] Food Composition and Methods Development Laboratory, Beltsville Human Nutrition Research Center, Agricultural Research Service, U.S. Department of Agriculture, MD 20705, USA; pei.chen@ars.usda.gov

* Correspondence: mnavarro@codeti.org; Tel.: +506-8873-5539

Received: 5 December 2017; Accepted: 22 January 2018; Published: 30 January 2018

Abstract: The phenolic composition of skin and flesh from *Malus domestica* apples (Anna cultivar) and *Prunus domestica* plums (satsuma cultivar) commercial cultivars in Costa Rica, was studied using Ultra Performance Liquid Chromatography coupled with High Resolution Mass Spectrometry (UPLC-DAD-ESI-MS) on enriched-phenolic extracts, with particular emphasis in proanthocyanidin and flavonoids characterization. A total of 52 compounds were identified, including 21 proanthocyanidins ([(+)-catechin and (−)-epicatechin]) flavan-3-ols monomers, five procyanidin B-type dimers and two procyanidin A-type dimers, five procyanidin B-type trimers and two procyanidin A-type trimers, as well as one procyanidin B-type tetramer, two procyanidin B-type pentamers, and two flavan-3-ol gallates); 15 flavonoids (kaempferol, quercetin and naringenin derivatives); nine phenolic acids (protochatechuic, caffeoylquinic, and hydroxycinnamic acid derivatives); five hydroxychalcones (phloretin and 3-hydroxyphloretin derivatives); and two isoprenoid glycosides (vomifoliol derivatives). These findings constitute the first report of such a high number and diversity of compounds in skins of one single plum cultivar and of the presence of proanthocyanidin pentamers in apple skins. Also, it is the first time that such a large number of glycosylated flavonoids and proanthocyanidins are reported in skins and flesh of a single plum cultivar. In addition, total phenolic content (TPC) was measured with high values observed for all samples, especially for fruits skins with a TPC of 619.6 and 640.3 mg gallic acid equivalents/g extract respectively for apple and plum. Antioxidant potential using 2,2-diphenyl-1-picrylhidrazyl (DPPH) and oxygen radical absorbance capacity (ORAC) methods were evaluated, with results showing also high values for all samples, especially again for fruit skins with IC_{50} of 4.54 and 5.19 µg/mL (DPPH) and 16.8 and 14.6 mmol TE/g (ORAC) respectively for apple and plum, indicating the potential value of these extracts. Significant negative correlation was found for both apple and plum samples between TPC and DPPH antioxidant values, especially for plum fruits ($R = -0.981$, $p < 0.05$) as well as significant positive correlation between TPC and ORAC, also especially for plum fruits ($R = 0.993$, $p < 0.05$) and between both, DPPH and ORAC antioxidant methods ($R = 0.994$, $p < 0.05$).

Keywords: *Malus domestica*; *Prunus domestica*; apple; plum; UPLC; ESI-MS; proanthocyanidins; flavonoids; mass spectrometry; antioxidant

---

1. Introduction

Polyphenols effects on health are based on results obtained from bioactivity studies, which in turn, has increased the interest in the consumption of foods and beverages rich in polyphenols as well as the importance of related scientific research. Such bioactive effects include antioxidant properties, prevention of oxidative stress associated diseases like cardiovascular, neurodegenerative diseases and cancer [1] and their role in long-term health protection by reducing the risk of chronic and degenerative diseases [2].

Several studies have linked vegetable consumption, specially fruits with a reduced risk for cancer and cardiovascular disease, thus characterization of polyphenols is essential to increase the knowledge on fruits contents and their related beneficial effects. *Malus domestica* (apple) and *Prunus domestica* (plum) are trees from the Rosaceae family, both native from southern Europe and western Asia that were introduced in Costa Rica as an initiative of local producers to diversify their crops, constitute fruits of high consumption in the country.

Studies on *M. domestica* have found these fruits to have a potent antioxidant activity and to inhibit the growth of cancer cells in vitro [3,4]. Similar properties have been described for plum [5,6] and some of these effects were attributed at least partially to its polyphenolic contents.

Previous studies on polyphenols of *M. domestica* have shown mainly flavonoids, phenolic acids, and chalcones, while proanthocyanidins included only catechin and epicatechin monomers and procyanidin dimers [7–9] and other reporting also trimers mainly in skin [10,11]. In the case of *P. domestica*, studies have focused mostly in a specific type of compounds, such as caffeolylquinic acid derivatives [12,13], flavonoids such as quercetin derivatives [14], and both type of compounds [15]. Few reports have studied also extensively proanthocyanidins, findings indicating mainly monomers and procyanidin dimers [16,17].

However other findings [18] would suggest proanthocyanidins with higher polymerization degree, without having reported their particular characterization. Despite the increasing number of studies on phenolics, the characterization of proanthocyanidins remains a complex task because of the need for high-end techniques such as high-resolution mass spectroscopy (HRMS). Proanthocyanidins, which are condensed flavan-3-ols, constitute an important group of polyphenols because of their bioactivities, for instance, mainly to their antioxidant capacity, which in turn is linked, among others, with their ant-inflammatory and anti-cancer activities [19].

Antioxidant properties have been reported to increase with proanthocyanidins degree of polymerization and the presence of specific structures such as procyanidin tetramers and pentamers has been found to enhance such properties [20,21], thus further knowledge on these phenolic structures characterization in apple and plum would contribute to a better understanding of their implications in the fruits quality as a source of dietary compounds with potential biological properties. Therefore, the objective of the present work was to obtain enriched polyphenolic extracts of fruits from *M. domestica* and *P. domestica* commercial cultivars in Costa Rica, and to characterize them through Ultra Performance Liquid Chromatography coupled with High Resolution Mass Spectrometry (UPLC-DAD-ESI-MS), with particular emphasis in flavonoids and proanthocyanidins. Evaluation of the total polyphenolic contents and antioxidant activity using 2,2-diphenyl-1-picrylhidrazyl (DPPH) and oxygen radical absorbance capacity (ORAC) methods, was also carried out in the different extracts.

2. Results and Discussion

2.1. Phenolic Yield and Total Phenolic Contents

The extraction process described in the Materials and Methods section, allowed to obtain phenolic enriched extracts, as summarized in Table 1. *P. domestica (satsuma cultivar)* skin presented the highest yield (2.74%) whereas *M. domestica (Anna cultivar)* flesh showed the lowest value (0.51%). In both fruits, skin extract yields were higher than flesh extracts. The total phenolic contents (TPC) shown also in Table 1, resulted in high values for all samples with both skins exhibiting slightly higher results, ranging from 619.6–640.3 gallic acid equivalents (GAE)/g dry extract.

Table 1. Extraction yield and total phenolic content.

Sample	Lyophilization Yield (%) [1]	Extraction Yield (%) [2]	Total Phenolic Content (mg GAE/g Extract) [3,4]
M. domestica			
Anna-Skin	11.6	1.20	619.6 [a,b] ± 19.5
Anna-Flesh	13.8	0.51	576.0 [a] ± 20.9
P. domestica			
Satsuma-Skin	13.1	2.74	640.3 [b] ± 22.7
Satsuma-Flesh	17.1	1.24	515.2 [c] ± 17.3

[1] g of dry material/g of fresh weight expressed as %. [2] g of extract/g of dry material expressed as %. [3] Values are expressed as mean ± Standard Deviation (S.D.). [4] Different superscript letters in the column indicate differences are significant at $p < 0.05$ using one-way analysis of variance (ANOVA) with a Tukey post hoc as statistical test. GAE: gallic acid equivalents.

Results from literature show variability among reports from other apple cultivars with total phenolic contents (TPC) values ranging between 0.3–25.9 mg GAE/g DW for skin and 1.6–14.8 mg GAE/g DW for flesh [22–25]. Our results for Anna cultivar skin and flesh 7.4 and 3 mg GAE/g DW respectively (values calculated using TPC and extract yields from Table 1) are within that range. A similar situation occurs in the case of plum, with determination of total phenolic contents in the literature revealing variability, with values ranging between 18.4–495 mg GAE/100 g fresh weight (FW) [14,26,27], whereas our findings of 109–179 mg GAE/100 g FW (values calculated using TPC, extract and lyophilization yields from Table 1) are in agreement with the published results.

2.2. Profile by UPLC-DAD-ESI-TQ-MS Analysis

The UPLC-DAD-ESI-MS/MS analysis described in the Materials and Methods section, allowed to identify 52 compounds, including 21 proanthocyanidins and flavan-3-ol monomers, 15 glycosylated flavonols and nine acids and derivatives in Costa Rican apple (Anna cultivar) and plum (Satsuma cultivar) only commercial cultivars; as well as five hydroxy chalcones and two glycosylated isoprenoids characteristic of apple fruits. Figures 1 and 2 show the chromatograms of the samples and Table 2 summarizes the analysis results for the 52 compounds.

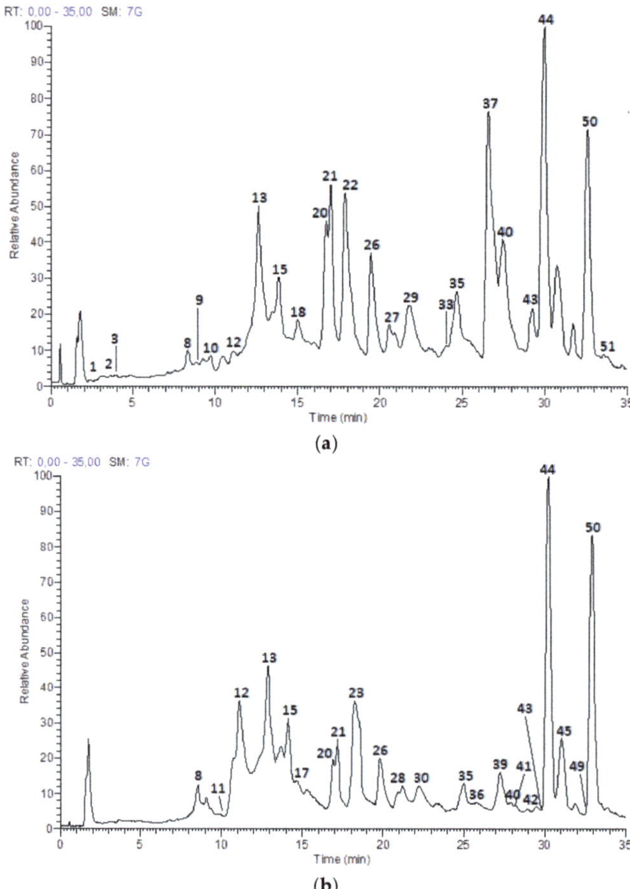

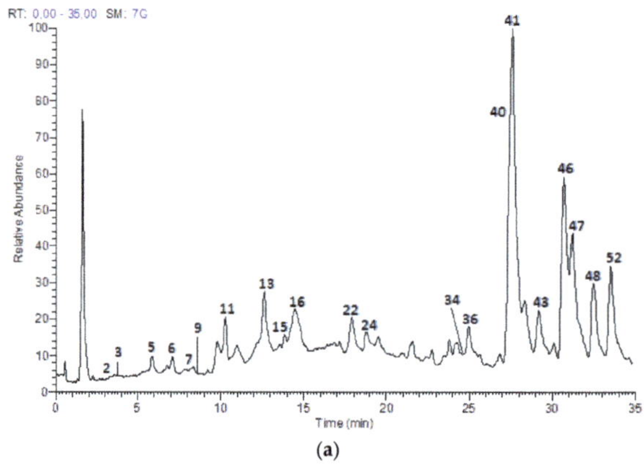

Figure 1. HPLC Chromatograms of *M. domestica* extracts: (**a**) Anna skins (**b**) Anna flesh, in a Hypersil Gold AQ RP-C18 column (200 mm × 2.1 mm × 1.9 μm) using a LTQ Orbitrap XL Mass spectrometer (Thermo Scientific™, Walthman, MA, USA) in a mass range from 100 to 2000 amu.

Figure 2. *Cont.*

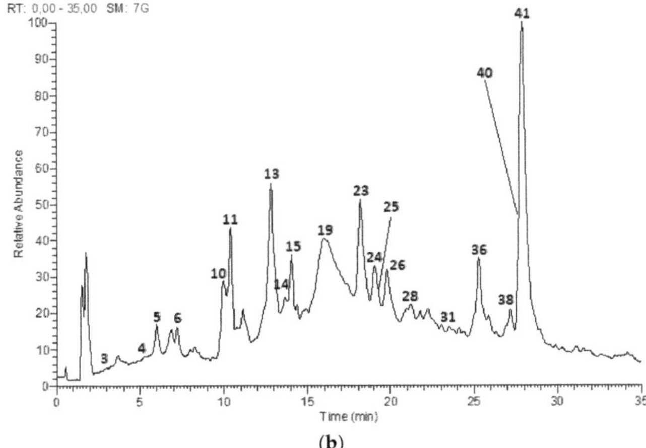

(b)

Figure 2. HPLC Chromatograms of *P. domestica* extracts: (**a**) Satsuma skin (**b**) Satsuma flesh in a Hypersil Gold AQ RP-C18 column (200 mm × 2.1 mm × 1.9 μm) using a LTQ Orbitrap XL Mass spectrometer (Thermo Scientific™, Walthman, MA, USA) in a mass range from 100 to 2000 amu.

Table 2. Profile of phenolic compounds identified by UPLC-DAD-ESI-TQ-MS analysis for apple and plum samples.

No.	Tentative Identification	t_R (min)	λ_{max} (nm)	$[M - H]^-$	Formula	MS^2 Fragments (% Abundance)	Apple Anna Skin	Apple Anna Flesh	Plum Satsuma Skin	Plum Satsuma Flesh
					Proanthocyanidins					
1	Procyanidin B-type dimer	2.69	277	577.1344	$C_{30}H_{26}O_{12}$	[577]: 289(28), 407(79), 425(100), 451(49), 559(66)	x			
3	(epi)catechin 3-O-gallate	3.96	284	441.0818	$C_{22}H_{18}O_{10}$	[441]: 153(31), 289(35), 315(100)	x		x	x
4	(epi)catechin 3-O-gallate	5.40	284	441.0819	$C_{22}H_{18}O_{10}$	[441]: 153(32), 289(28), 315(100)				x
8	Procyanidin B-type dimer	8.44	279	577.1344	$C_{30}H_{26}O_{12}$	[577]: 289(50), 407(70), 425(100), 451(80), 559(42)	x	x		
9	Catechin	8.87	289	289.0709	$C_{15}H_{20}O_6$	[289]: 205(38),245(100)	x		x	
13	Procyanidin B-type dimer	12.75	282	577.1344	$C_{30}H_{26}O_{12}$	[577]: 289(35), 407(57), 425(100), 451(56), 559(28)	x	x	x	x
14	Procyanidin B-type trimer	13.20	282	865.1956	$C_{45}H_{38}O_{18}$	[865]: 577(43), 695(100), 713(39), 739(60)				x
15	Epicatechin	13.97	280	289.0707	$C_{15}H_{14}O_6$	[289]: 205(35), 245(100)	x	x	x	x
22	Procyanidin B-type trimer	17.88	279	865.1956	$C_{45}H_{38}O_{18}$	[865]: 577(54), 695(100), 713(37), 739(71)	x		x	
23	Procyanidin B-type trimer	18.16	278	865.1956	$C_{45}H_{38}O_{18}$	[865]: 577(61), 695(100), 713(33), 739(67)		x		x
24	Procyanidin A-type trimer	18.98	278	863.1798	$C_{45}H_{36}O_{18}$	[863]: 575(100), 711(63)			x	x
25	Procyanidin A-type trimer	19.37	277, 517	863.1798	$C_{45}H_{36}O_{18}$	[863]: 575(100), 711(63)				x
26	Procyanidin B-type tetramer	19.70	279, 517	1153.2603	$C_{60}H_{50}O_{24}$	[1153]: 575(43), 577(46), 863(62), 865(100), 983(87), 1001(37), 1027(66)	x	x		x
27	Procyanidin B-type pentamer	20.54	280	1441.3229	$C_{75}H_{62}O_{30}$	[1441]: 1315(43), 1151(70), 863(68), 635(100), 577(40)	x			
28	Procyanidin B-type trimer	21.23	279	865.1956	$C_{45}H_{38}O_{18}$	[865]: 407(45), 577(59), 695(100), 713(66), 739(73)		x		x
29	Procyanidin B-type pentamer	21.83	278	1441.3229	$C_{75}H_{62}O_{30}$	[1441]: 1315(33), 1151(69), 863(95), 635(100), 577(60)	x			
30	Procyanidin B-type trimer	22.21	279	865.1956	$C_{45}H_{38}O_{18}$	[865]: 575(46), 577(53), 695(100), 713(44), 739(84)		x		
31	Procyanidin A-type trimer	23.54	282	863.1798	$C_{45}H_{36}O_{18}$	[863]: 575(100), 711(58)				x
36	Procyanidin A-type dimer	25.19	279	575.1185	$C_{30}H_{24}O_{12}$	[575]: 289(35), 449(100)		x	x	x
38	Procyanidin B-type dimer	27.21	276	577.1344	$C_{30}H_{26}O_{12}$	[577]: 289(34), 407(60), 425(100), 451(78), 559(31),				x
42	Procyanidin B-type dimer	28.62	279	577.1344	$C_{30}H_{26}O_{12}$	[577]: 289(50), 407(65), 425(100), 451(76), 559(44)		x		

Table 2. Cont.

No.	Tentative Identification	t_R (min)	λ_{max} (nm)	$[M − H]^-$	Formula	MS^2 Fragments (% Abundance)	Apple Anna Skin	Apple Anna Flesh	Plum Satsuma Skin	Plum Satsuma Flesh
					Glycosylated flavonols					
18	Kaempferol-hexoside	14.97	280, 351	447.0922	$C_{21}H_{20}O_{11}$	[447]: 284(70), 285(100)	x			
19	Kaempferol-hexoside	15.90	278, 360, 516	447.0922	$C_{21}H_{20}O_{11}$	[447]: 284(23), 285(100)				x
32	Naringenin-hexoside	23.97	278, 351	433.1131	$C_{21}H_{22}O_{10}$	[433]: 271(100)	x			
33	Quercetin-pentosylhexoside	24.12	279, 351	595.1284	$C_{26}H_{28}O_{16}$	[595]: 300(100), 301(41)	x			
34	Quercetin-pentosylhexoside	24.72	281, 357	595.1284	$C_{26}H_{28}O_{16}$	[595]: 300(100), 301(40)			x	
37	Quercetin-hexoside	26.57	255, 350	463.0875	$C_{21}H_{20}O_{12}$	[463]: 300(36), 301(100)	x			
40	Quercetin-rutinoside	27.60	255, 360	609.1440	$C_{27}H_{30}O_{16}$	[609]: 300(31), 301(100)	x	x	x	x
41	Quercetin-hexoside	27.79	252, 351	463.0875	$C_{21}H_{20}O_{12}$	[463]: 300(28), 301(100)		x	x	x
43	Quercetin-pentoside	29.28	258, 355	433.0769	$C_{20}H_{18}O_{11}$	[433]: 300(29), 301(100)	x	x	x	
46	Quercetin-pentoside	30.69	258, 347	433.0769	$C_{20}H_{18}O_{11}$	[433]: 301(100)			x	
47	Quercetin-pentosylpentoside	31.21	258, 354	565.1184	$C_{25}H_{26}O_{15}$	[565]: 300(100), 301(16)			x	
48	Quercetin-deoxyhexoside	32.46	355	447.0922	$C_{21}H_{20}O_{11}$	[447]: 300(26), 301(100)			x	
49	Quercetin-deoxyhexoside	32.57	284	447.0922	$C_{21}H_{20}O_{11}$	[447]: 300(30), 301(100)		x		
51	Quercetin-deoxyhexoside	33.27	281	447.0922	$C_{21}H_{20}O_{11}$	[447]: 300(30), 301(100)	x			
52	Quercetin-acetylhexoside	33.51	354	505.0975	$C_{23}H_{22}O_{13}$	[505]: 300(63), 301(100)			x	
					Acids and derivates					
2	Protocatechuic acid	3.36	280	153.0191	$C_7H_6O_4$	[153]: 109(100)	x		x	
5	Caffeoylquinic acid isomer	5.95	323	353.0869	$C_{16}H_{18}O_9$	[353]: 191(100), 179(71)			x	x
6	Caffeoyl hexoside	7.23	331	341.0872	$C_{15}H_{18}O_9$	[341]: 161(37), 179(100)			x	x
7	Coumaric acid	8.30	313	163.0398	$C_9H_6O_3$	[163]: 119(100)			x	
10	p-coumaroyl-hexoside	9.94	297	325.0921	$C_{15}H_{18}O_8$	[325]: 145(100), 163(92), 187(49)	x			x
11	p-coumaroyl-hexoside	10.27	314	325.0921	$C_{15}H_{18}O_8$	[325]: 145(100), 163(87), 187(50)		x	x	x
12	Caffeoylquinic acid isomer	11.10	270, 313	353.0869	$C_{16}H_{19}O_9$	[353]: 191(100), 145(46)	x	x		
16	Shikimic acid	14.44	316	173.0454	$C_7H_{10}O_5$	[173]: 93(100),111(43)			x	
17	p-coumaroylquinic acid	14.49	311	337.0927	$C_{16}H_{18}O_8$	[337]: 173(100)		x		
					Chalcones					
35	3-hydroxyphloretin-pentosylhexoside	24.87	281	583.1660	$C_{26}H_{32}O_{15}$	[583]: 289(100)	x	x		
39	3-hydroxyphloretin	27.29	283	289.0716	$C_{15}H_{14}O_6$	[289]: 167(100), 245(49), 271(81)		x		
44	Phloretin-pentosilhexoside	30.11	284	567.1704	$C_{26}H_{32}O_{14}$	[567]: 273(100)	x	x		
45	Phloretin-pentosilhexoside	31.04	283	567.1704	$C_{26}H_{32}O_{14}$	[567]: 273(100)		x		
50	Phloretin	32.78	283	273.0767	$C_{15}H_{14}O_5$	[273]: 167(100)	x	x		
					Other compounds					
20	Vomifoliol-pentosilhexoside	16.80	281	517.2280	$C_{24}H_{38}O_{12}$	[517]: 205(100), 385(58)	x	x		
21	Vomifoliol-pentosilhexoside	17.10	281	517.2280	$C_{24}H_{38}O_{12}$	[517]: 205(100), 385(64)	x	x		

The first common group of compounds, proanthocyanidins, corresponded to oligomers of flavan-3-ols catechin and epicatechin. The monomeric units of these proanthocyanidins are linked through a C4-C8 or C4-C6 bond (B-type), which coexist with an additional C2-O-C7 linkage (A-type) [28]. Peaks 9 (Rt = 8.87 min) and 15 (Rt = 13.97 min) showed a $[M - H]^-$ at m/z 289.0710 ($C_{15}H_{14}O_6$) that correspond to monomers catechin or epicatechin. The main MS^2 fragments at m/z 245 and 205, occur through the loss of C_2H_4O and $C_4H_4O_2$ due to retro-Diels-Alder fission (RDA) of ring A [29].

On the other hand, peaks 3 (Rt = 3.96 min), 4 (Rt = 5.40 min), whose $[M - H]^-$ is at m/z 441.0819 ($C_{22}H_{18}O_{10}$) correspond to (epi)catechin-3-O-gallate (Figure 3). The main fragment at m/z 315 $[M - H - 126]^-$ is due to the elimination of a phloroglucinol, and fragments at m/z 289 and 153 are both residuals from the cleavage of the ester group [30].

Compound No.	R
3, 4	Galloyl
9, 15	H

Figure 3. Flavan-3-ols monomers and gallates structure and main fragments.

Peak 36 (Rt = 25.19 min) shows a $[M - H]^-$ at m/z 575.1185 ($C_{30}H_{24}O_{12}$) and main MS^2 ion at m/z 449, which indicate the presence of a procyanidin A-type dimer. The base ion at m/z 449 $[M - H - 126]^-$, corresponds to the elimination of a phloroglucinol molecule from this A-type dimer [31]. Peaks 24 (Rt = 18.98 min), 25 (Rt = 19.37 min) and 31 (Rt = 23.54 min) show a $[M - H]^-$ at m/z 863.1798 ($C_{45}H_{36}O_{18}$), revealing the presence of a procyanidin trimer with A-type interflavan linkage (Figure 4). In the MS^2 spectrum, fragment ions at m/z 711 $[M - H - 152]^-$ and 575 $[M - H - 288]^-$ observed, result from the RDA fission and quinone-methide (QM) cleavage, respectively [32].

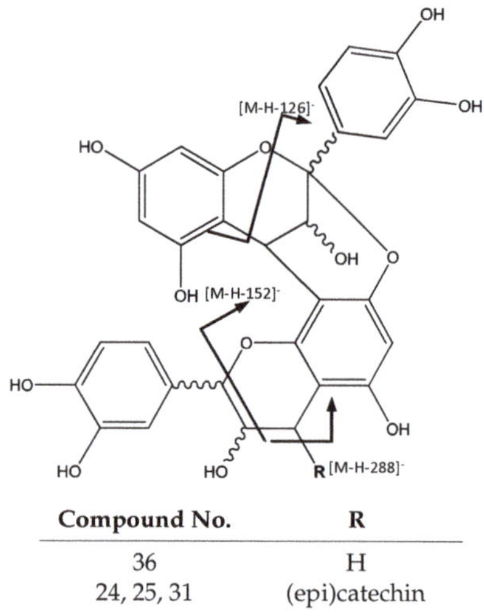

Compound No.	R
36	H
24, 25, 31	(epi)catechin

Figure 4. Proanthocyanidin A-type structure and main fragments.

Peaks 1 (Rt = 2.69 min), 8 (Rt = 8.44 min), 13 (Rt = 12.75 min), 38 (Rt = 27.21 min) and 42 (Rt = 28.62 min) show $[M - H]^-$ at m/z 577.1344 ($C_{30}H_{26}O_{12}$), corresponding to procyanidins with B-type linkage (Figure 5), 2 amu (atomic mass units) higher than that of the A-type procyanidin, and major ions containing the structural information at m/z 559, 451, 425, 407 and 289. The ion at m/z 559 $[M - H - 18]^-$ originates from water loss. The ion at m/z 451 $[M - 126 - H]^-$ results from the elimination of the phloroglucinal as in A-type dimers. The fragment ions at m/z 425 $[M - H - 152]^-$ and 407 $[M - H - 170]^-$ come from RDA, while the ion at m/z 289 originates from QM resulting in the ion of the monomer [31].

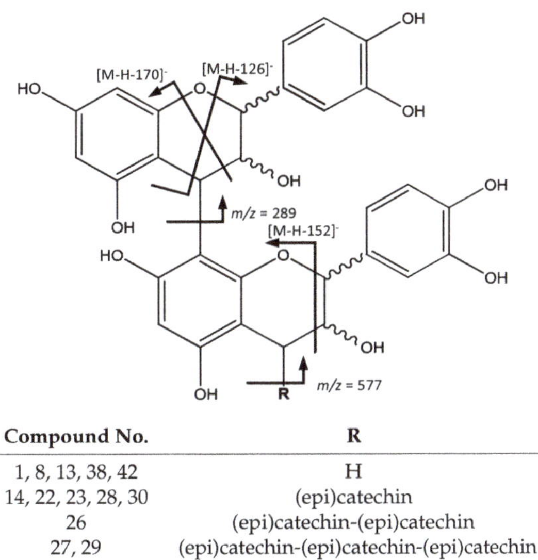

Compound No.	R
1, 8, 13, 38, 42	H
14, 22, 23, 28, 30	(epi)catechin
26	(epi)catechin-(epi)catechin
27, 29	(epi)catechin-(epi)catechin-(epi)catechin

Figure 5. Proanthocyanidin B-type structure and main fragments.

On the other hand (Figure 5), peaks 14 (Rt = 13.20 min), 22 (Rt = 17.88 min), 23 (Rt = 18.16 min), 28 (Rt = 21.23 min) and 30 (Rt = 22.21 min) with ($[M - H]^-$) at m/z 865.1956 ($C_{45}H_{38}O_{18}$) were tentatively identified to be procyanidin B-type trimers. Their fragmentation behaviors seem to be similar to that of dimers, with ion fragments at m/z 695 $[M - H - 170]^-$, 713 $[M - H - 152]^-$ and 739 $[M - H - 126]^-$. The QM cleavage of the interflavan bond mainly produced the ions at m/z 289 and 577, indicating the cleavage happens in upper interflavan bond [31].

In a similar way (Figure 5), peak 26 (Rt = 19.70 min), $[M - H]^-$ at m/z 1153.2603 ($C_{60}H_{50}O_{24}$) was identified as a procyanidin B-type tetramer, with fragments at m/z 1027 $[M - H - 126]^-$, 1001 $[M - H - 152]^-$, 983 $[M - H - 170]^-$, 865 and 577. Also, peaks 27 (Rt = 20.54 min) and 29 (Rt = 21.83 min), with $[M - H]^-$ at m/z 1441.3229 ($C_{75}H_{62}O_{30}$), were identified as two procyanidin B-type pentamers with a characteristic fragment at m/z 1315 $[M - H - 126]^-$, and also those derived from QM cleavage at m/z 1151, 865, 577 and 289 [33].

The second group of common compounds, glycosylated flavonol derivatives were elucidated based in the fragmentation pattern from the aglycone due to the loss of glycosides (Figure 6). For instance, peaks 18 (Rt = 14.97 min) and 19 (Rt = 15.90 min) had $[M - H]^-$ at m/z 447.0922 ($C_{21}H_{20}O_{11}$) were identified as kaempferol-hexoside isomers with a main fragment at m/z 285 corresponding to kaempferol [34]. Peak 32 (Rt = 23.97 min) with $[M - H]^-$ at m/z 433.1131 ($C_{21}H_{22}O_{10}$) showed its main fragment at m/z 271, corresponding to naringenin-hexoside [35].

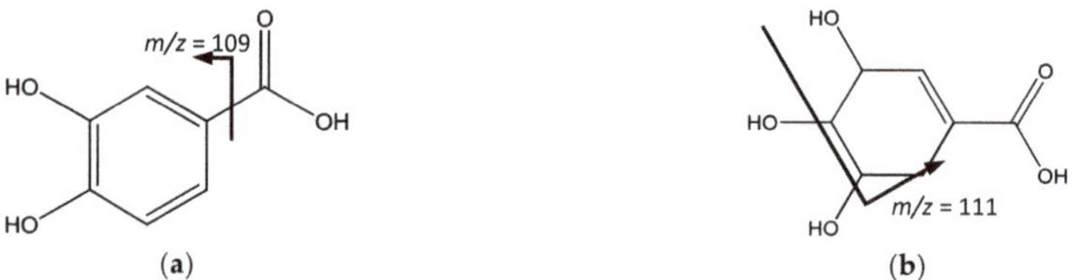

Compound No.	O-R$_1$	R$_2$	R$_3$
18, 19	O-hexoside	H	H
32	H	H	hexoside
33, 34	O-pentosylhexoside	OH	H
37, 41	O-hexoside	OH	H
40	O-rutinoside	OH	H
43, 46	O-pentoside	OH	H
47	O-pentosylpentoside	OH	H
48, 49, 51	O-deoxyhexoside	OH	H
52	O-acetylhexoside	OH	H

Figure 6. Flavonol glycosides structure and main fragments.

All the remaining flavonol derivatives presented their main fragment at m/z 301, which corresponds to the quercetin aglycone. They differ in the bonded glycoside with some variations among them. For instance, peaks 33 (Rt = 24.12 min) and 34 (Rt = 24.72 min) had a [M − H]⁻ at m/z 595.1284 ($C_{26}H_{28}O_{16}$), were assigned to quercetin-pentosylhexoside isomers. Peaks 37 (Rt = 26.57 min) and 41 (Rt = 27.79 min), with [M − H]⁻ at m/z 463.0875 ($C_{21}H_{20}O_{12}$) correspond to quercetin-hexoside isomers. Peak 40 (Rt = 27.60 min), with [M − H]⁻ at m/z 609.1440 ($C_{27}H_{30}O_{16}$) was identified as quercetin-rutinoside. Peaks 43 (Rt = 29.28 min) and 46 (Rt = 30.69 min) had [M − H]⁻ at m/z 433.0769 ($C_{20}H_{18}O_{11}$), coincident with isomers of quercetin-pentoside [36].

Peak 47 (Rt = 31.21 min) had [M − H]⁻ at m/z 565.1184 ($C_{25}H_{26}O_{15}$), were identified as quercetin-pentosyl-pentoside. Peaks 48 (Rt = 32.46 min), 49 (Rt = 32.57 min) and 51 (Rt = 33.27 min) with [M − H]⁻ at m/z 447.0922 ($C_{21}H_{20}O_{11}$) were assigned as quercetin-deoxyhexoside isomers. Finally, peak 52 (Rt = 33.51 min) with [M − H]⁻ at m/z 505.0975 ($C_{23}H_{22}O_{13}$), was identified as quercetin-acetylhexoside [16].

Among the third group of common compounds, acids and derivatives, two small acids correspond (Figure 7) to peak 2 (Rt = 3.36 min) with [M − H]⁻ at m/z 153.0191 ($C_7H_6O_4$) and a main fragment at m/z 109 [M – H − 44]⁻ due to the loss of CO_2 from a carboxylic acid [37] identified as protocatechuic acid, and peak 16 (Rt = 14.44 min), with [M − H]⁻ at m/z 173.0454 ($C_7H_{10}O_5$) and main fragments at m/z 111 generated from RDA fission, and 93 from subsequent loss of water assigned to shikimic acid [38].

(a) (b)

Figure 7. (**a**) Protochatechuic acid and (**b**) Shikimic acid structure and main fragments.

On the other hand, a series of *p*-coumaric acid derivatives was identified, as shown in Figure 8. For instance, peaks 5 (Rt = 5.95 min) and 12 (Rt = 11.10 min), with [M − H]⁻ at m/z 353.0869 ($C_{16}H_{18}O_9$), are identified as caffeoylquinic acid isomers, with main fragments at m/z 191 [quinic acid$-$H]$^-$, 179 [caffeic acid$-$H]$^-$, and 145 due to the loss of CO_2 from the quinic acid ion. Peak 6 (Rt = 7.23 min) shows [M $-$ H]$^-$ at m/z 341.0872 ($C_{15}H_{18}O_9$), with main fragments at m/z 179 [caffeic acid $-$ H]$^-$ and 161 [M $-$ H $-$ 179]$^-$, corresponding to caffeoyl-hexoside. Peak 7 (Rt = 8.30 min) with [M $-$ H]$^-$ at m/z 163.0398 ($C_9H_6O_3$), is identified as coumaric acid due to the fragment at 119 [M $-$ H $-$ CO_2]$^-$. Peaks 10 (Rt = 9.94 min) and 11 (Rt = 10.27 min), with [M $-$ H]$^-$ at m/z 325.0921 ($C_{15}H_{18}O_8$) are assigned to coumaroyl-hexoside isomers, due to fragments at m/z 163 [coumaric acid$-$H]$^-$ and 145 [coumaric acid$-$H$-$H$_2$O]$^-$. Another cinnamic acid derivative was found in peak 17 (Rt = 14.49 min) with [M $-$ H]$^-$ at m/z 337.0927 ($C_{16}H_{18}O_8$), and a main fragment at m/z 173 due to the loss of water of the quinic acid ion, thus corresponding to *p*-coumaroylquinic acid [16].

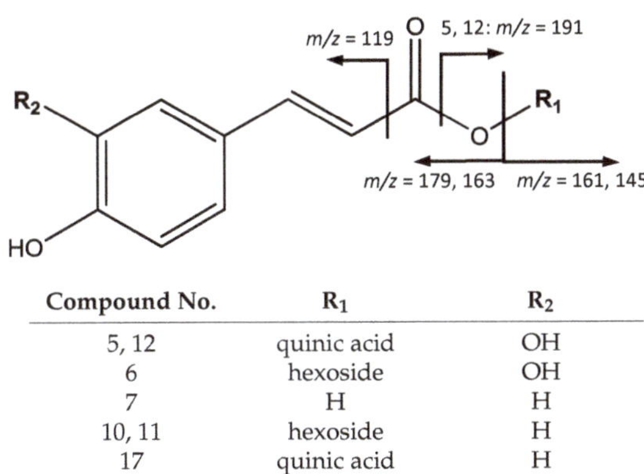

Compound No.	R_1	R_2
5, 12	quinic acid	OH
6	hexoside	OH
7	H	H
10, 11	hexoside	H
17	quinic acid	H

Figure 8. *p*-coumaric acid derivatives structure and main fragments.

The fourth group of compounds, chalcones, shown in Figure 9, were found only in apples. For instance, peak 35 (Rt = 24.87 min) shows [M $-$ H]$^-$ at m/z 583.1660 ($C_{26}H_{32}O_{15}$) and the main fragment at m/z 289, which correspond to 3-hydroxyphloretin aglycone, allowing to identify the compound as 3-hydroxyphloretin-pentosylhexoside [11]. Peak 39 (Rt = 27.29 min) shows [M $-$ H]$^-$ at m/z 289.0716 ($C_{15}H_{14}O_6$), the same mass and molecular formula of flavan-3-ols, but differs in the fragmentation that occurs at m/z 271 [M $-$ H $-$ H$_2$O]$^-$ due to the loss of water, 245 coming from RDA, and 167 due to α-cleavage of the carbonyl group [39] therefore being assigned as 3-hydroxyphloretin.

On the other hand, peaks 44 (Rt = 30.11 min) and 45 (Rt = 31.04 min) with [M $-$ H]$^-$ at m/z 567.1704 ($C_{26}H_{32}O_{14}$) were identified as phloretin-pentosylhexoside isomers, due to the main fragment at m/z 273 which corresponds to the phloretin ion generated from loss of glycosides. Finally, peak 50 (Rt = 32.78 min), [M $-$ H]$^-$ at m/z 273.0767 ($C_{15}H_{14}O_5$) is assigned as phloretin, with a main fragment at m/z 167 due to α-cleavage of the carbonyl group [11,39].

The fifth group of compounds, glycosylated isoprenoid derivatives, shown in Figure 10, was found only in apple. Peaks 20 (Rt = 16.80 min) and 21 (Rt = 17.10 min) showed [M $-$ H]$^-$ at m/z 517.2280 ($C_{24}H_{38}O_{12}$). Their main fragments at m/z 385 [M $-$ H $-$ 132]$^-$ and 205 [M $-$ H $-$ 312]$^-$ correspond to the loss of a pentoside and a pentosylhexoside respectively. The resulting ion is coincident with vomifoliol, allowing the peaks to be assigned to vomifoliol-pentosylhexoside isomers [40].

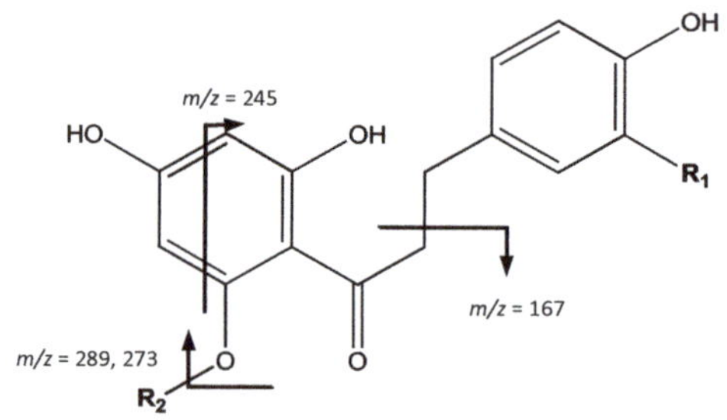

Compound No.	R₁	R₂
35	OH	pentosylhexoside
39	OH	H
44, 45	H	Pentosylhexoside
50	H	H

Figure 9. Chalcones structure and main fragments.

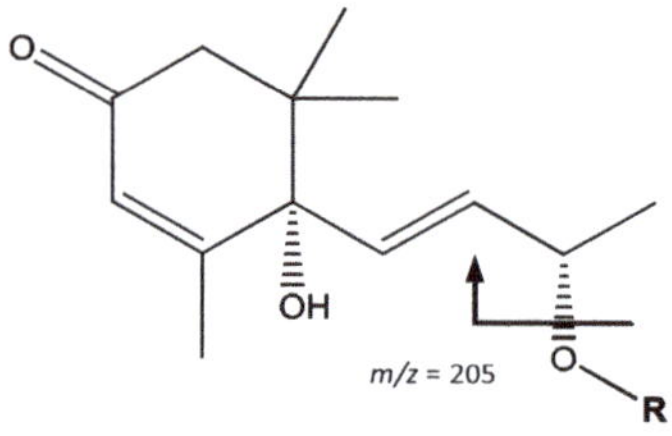

Compound No.	R
20, 21	pentosylhexoside

Figure 10. Vomifoliol-pentosylhexoside structure and main fragments.

Polyphenol profiling reveals great similarities between skins and flesh, with a high number of compounds and high diversity for both fruits. When comparing the reports for apple cultivars in the literature, our results for Anna cultivar from Costa Rica show greater number and diversity of polyphenols than the findings on sixteen cultivars from Norway, Italy, Canada and United States [7–10], and similar to only two cultivars, namely Golden Delicious and Braeburn from Slovenia [11]. Further, findings on proanthocyanidins indicate better results both in total occurrence and in greater polymerization degree, for instance procyanidin tetramers and pentamers in the skins of Costa Rican Anna apple cultivar.

In the case of plums, when comparing the reports of twenty-four cultivars from United States, Germany and Portugal [15–18], greater number and diversity of polyphenols is observed in the skins of Satsuma cultivar from Costa Rica. Likewise, results for Satsuma flesh are also superior to twenty-three of the cultivars, being similar to President cultivar from Germany. Of special interest is the presence of a greater number of procyanidins oligomers, such as procyanidin trimers and one tetramer as well as of glycosylated flavonols, showing different quercetin derivatives not reported in previous characterizations of plum skins.

It is important to highlight the characterization of proanthocyanidin oligomers present in both fruits, which is of special interest due to procyanidins antioxidant results having been reported to increase with higher degree of polymerization, for instance trimers, tetramers and pentamers showing better results than dimers or flavan-3-ol monomers [20,21]. The detailed characterization of these fruit cultivars provides an important contribution expanding the knowledge for their exploitation as available sources of such diversity of polyphenols, which in turn can be of interest for further research due to their potential biological activities.

2.3. Antioxidant Activity

The DPPH and ORAC values obtained are summarized in Table 3. All samples show high antioxidant values and in both antioxidant tests, skins have better results than flesh, with Anna skin presenting the highest value with IC_{50} = 4.54 µg/mL for DPPH and 16.78 mmol Trolox equivalents/g for ORAC. Regarding antioxidant values from the literature, while no comparable results are available for plum fruits, evaluation of methanolic and aqueous extracts from Kahsmir cultivar apple skins show DPPH IC_{50} values of 55.54 µg/mL and 41.41 µg/mL respectively [41], thus Anna cultivar extracts showing better results. DPPH values expressed as mmol TE/g extract were obtained as described in the experimental section (in respect to Trolox IC_{50} = 5.62 µg/mL) and allow to compare our results with those reported in the literature for different extracts, for instance our values fit in the range between enju and grape seed extracts (0.76–1.35 mmol TE/g extract) used as antioxidant food additives [42,43].

Table 3. DPPH and ORAC antioxidant activity.

Sample	DPPH [1,2]		ORAC [1,2]
	IC_{50} (µg/mL)	(mmol TE/g Extract)	(mmol TE/g Extract)
M. domestica			
Anna-Skin	4.54 [a] ± 0.06	1.24 [a] ± 0.02	16.78 [a] ± 0.25
Anna-Flesh	6.64 [b] ± 0.12	0.85 [b] ± 0.01	11.22 [b] ± 0.13
P. domestica			
Satsuma-Skin	5.19 [c] ± 0.12	1.08 [c] ± 0.02	14.55 [c] ± 0.21
Satsuma-Flesh	5.95 [d] ± 0.14	0.94 [d] ± 0.03	13.02 [d] ± 0.29

[1] Values are expressed as mean ± S.D. [2] Different superscript letters in the same column indicate differences are significant at $p < 0.05$ using ANOVA with a Tukey post hoc as statistical test. ORAC: oxygen radical absorbance capacity; DPPH: 2,2-diphenyl-1-picrylhidrazyl method.

On the other hand, for ORAC, a study of ethanolic extracts of Pelingo cultivar apples [10], reported values of 44.07 µmol Trolox eq./g DW for skin and 23.19 µmol Trolox eq./g DW for flesh, while findings for aqueous extracts showed values of 42.97 µmol Trolox eq./g DW and 31.99 µmol Trolox eq./g DW for skin and flesh respectively, therefore ORAC of extracts from Anna cultivar apples are superior for both skin and flesh since our findings indicate values of 57.33 µmol Trolox eq./g DW for flesh and 202.03 µmol Trolox eq./g DW (values calculated using ORAC from Table 3 and extract yields from Table 1).

The difference in antioxidant values among extracts could be attributed to the differences in their phenolic content and distribution. Thus, in order to investigate if the total phenolic contents (TPC, Table 1) contributes to the antioxidant activity, a correlation analysis was carried out between these TPC values with DPPH and ORAC results. Significant positive correlation ($p < 0.05$) was found for both apple and plum samples between TPC values and ORAC with R = 0.827 and R = 0.993 respectively, as well as significant negative correlation ($p < 0.05$) between TPC results and DPPH with $R = -0.833$ and R = −0.981 respectively. Therefore, our results are in agreement with previous studies reporting correlation between total polyphenolic contents and ORAC antioxidant activity [44]. Finally, our findings indicate positive correlation ($p < 0.05$) between both DPPH and ORAC antioxidant values (R = 0.994), thus in agreement with previous studies [43].

3. Materials and Methods

3.1. Materials, Reagents and Solvents

Malus domestica and *Prunus domestica* fruits were acquired in ripe state from FRUTALCOOP, a local producer cooperative located in Los Santos in Costa Rica. Cultivars were confirmed with the support of the Costa Rican National Herbarium and vouchers are deposited there. Reagents, such as fluorescein, 2,2-azobis(2-amidinopropane) dihydrochloride (AAPH), 2,2-diphenyl-1-picrylhidrazyl (DPPH), Trolox, gallic acid, and Amberlite XAD-7 resin were provided by Sigma-Aldrich (St. Louis, MO, USA), while solvents such as acetone, chloroform and methanol were purchased from Baker (Center Valley, PA, USA).

3.2. Phenolic Extracts from M. domestica and P. domestica Fruits

M. domestica and *P. domestica* fruits were rinsed in water, peeled, and both, skin and flesh material were freeze-dried in a Free Zone −105 °C, 4.5 L, Cascade Benchtop Freeze Dry System (Labconco, Kansas, MO, USA), and the freeze-dried material was preserved at −20 °C until extraction. Freeze-dried samples were extracted in a Dionex™ ASE™ 150 Accelerated Solvent Extractor (Thermo Scientific™, Walthman, MA, USA) using acetone:water (70:30) as solvent in a 34 mL cell, at 40 °C. Next, the extract was evaporated under vacuum to eliminate the acetone and the aqueous phase was washed with ethyl acetate and chloroform to remove less-polar compounds. Afterwards, the aqueous extract was evaporated under vacuum to eliminate organic solvent residues and was eluted (2 mL/min) in Amberlite XAD7 column (150 mm × 20 mm), starting with 300 mL of water to remove sugars, and then with 200 mL each of methanol:water (80:20) and pure methanol to obtain the polyphenols. Finally, the enriched extract was obtained after evaporating to dryness using a Buchi™ 215 (Flawil, Switzerland) rotavapor.

3.3. Total Phenolic Content

The polyphenolic content was determined by a modification of the Folin-Ciocalteu (FC) method [45], whose reagent is composed of a mixture of phosphotungstic and phosphomolybdic acids. Each sample was dissolved in MeOH (0.1% HCl) and combined with 0.5 mL of FC reagent. Afterwards 10 mL of Na_2CO_3 (7.5%) were added and the volume was completed to 25 mL with water. Blanks were prepared in a similar way but using 0.5 mL of MeOH (0.1% HCl) instead of sample. The mixture was let standing in the dark for 1 h and then absorbance was measured at 750 nm. Values obtained were extrapolated in a gallic acid calibration curve. Total phenolic content was expressed as mg gallic acid equivalents (GAE)/g sample. Analyses were performed in triplicate.

3.4. UPLC-DAD-ESI-TQ-MS Analysis

The UPLC-MS system used to analyze the composition of *M. domestica* and *P. domestica* fruit extracts consisted of an LTQ Orbitrap XL mass spectrometer with an Accela 1250 binary Pump, a PAL HTC Accela TMO autosampler, a PDA detector (Thermo Fisher Scientific, San Jose, CA, USA), and a G1316A column compartment (Agilent, Palo Alto, CA, USA). Separation was carried out on a Hypersil Gold AQ RP-C18 UHPLC column (200 mm × 2.1 mm i.d., 1.9 μm, Thermo Fisher Scientific) with an UltraShield pre-column filter (Analytical Scientific Instruments, Richmond, CA, USA) at a flow rate of 0.3 mL/min. Mobile phases A and B consist of a combination of 0.1% formic acid in water, *v/v* and 0.1% formic acid in acetonitrile, *v/v*, respectively. The linear gradient is from 4% to 20% B (*v/v*) at 20 min, to 35% B at 30 min and to 100% B at 31 min, and held at 100% B to 35 min. The UV/Vis spectra were acquired from 200–700 nm.

Negative electrospray ionization mode was used and the conditions were set as follows: sheath gas, 70 (arbitrary units); aux and sweep gas, 15 (arbitrary units); spray voltage, 4.8 kV; capillary temperature, 300 °C; capillary voltage, 15 V; tube lens, 70 V. The mass range was from 100 to 2000 amu with a resolution of 30,000, FTMS AGC target at 2×10^5, FT- MS/MS AGC target at 1×10^5, isolation width of 1.5 *amu*, and max ion injection time of 500 ms. The most intense ion was selected for the data-dependent scan to offer their MS^2 to MS^5 product ions, respectively, with a normalization collision energy at 35%.

3.5. DPPH Radical-Scavenging Activity

DPPH evaluation was performed as previously reported [46] and was expressed as IC_{50} (μg/mL), which is the amount of sample required to reach the 50% radical-scavenging activity, and also as mmol of Trolox equivalents (TE)/g extract. Briefly, a solution of 2,2-diphenyl-1-picrylhidrazyl (DPPH) (0.25 mM) was prepared using methanol as solvent. Next, 0.5 mL of this solution were mixed with 1 mL of extract or Trolox at different concentrations, and incubated at 25 °C in the dark for 30 min. DPPH absorbance was measured at 517 nm. Blanks were prepared for each concentration. The percentage of the radical-scavenging activity of the sample or Trolox was plotted against its concentration to calculate IC_{50} (μg/mL). The samples were analyzed in three independent assays. In order to express the DPPH results as mmol TE/g extract, the IC_{50} (μg/mL) of Trolox was converted to mmol/mL using Trolox molecular weight (250.29 mg/mmol) and then dividing by the IC_{50} of each sample.

3.6. ORAC Antioxidant Activity

The ORAC (Oxygen Radical Absorbance Capacity) antioxidant activity was determined following a method previously described [47] using fluorescein as a fluorescence probe. The reaction was performed in 75 mM phosphate buffer (pH 7.4) at 37 °C. The final assay mixture consisted of AAPH (12 mM), fluorescein (70 nM), and either Trolox (1–8 μM) or the extract at different concentrations. Fluorescence was recorded every minute for 98 min in black 96-well untreated microplates (Nunc, Denmark), using a Polarstar Galaxy plate reader (BMG Labtechnologies GmbH, Offenburg, Germany) with 485-P excitation and 520-P emission filters. Fluostar Galaxy software version 4.11-0 (BMG Labtechnologies GmbH, Offenburg, Germany) was used to measure fluorescence. Fluorescein was diluted from a stock solution (1.17 mM) in 75 mM phosphate buffer (pH 7.4), while AAPH and Trolox solutions were freshly prepared. All reaction mixtures were prepared in duplicate and three independent runs were completed for each extract. Fluorescence measurements were normalized to the curve of the blank (no antioxidant). From the normalized curves, the area under the fluorescence decay curve (AUC) was calculated as:

$$AUC = 1 + \sum_{i=1}^{i=98} f_i \Big/ f_0$$

where f_0 is the initial fluorescence reading at 0 min and f_i is the fluorescence reading at time i. The net AUC corresponding to a sample was calculated as follows:

$$Net\ AUC = AUC_{antioxidant} - AUC_{blank}$$

The regression equation between net AUC and antioxidant concentration was calculated. The ORAC value was estimated by dividing the slope of the latter equation by the slope of the Trolox line obtained for the same assay. Final ORAC values were expressed as mmol of Trolox equivalents (TE)/g of phenolic extract.

3.7. Statistical Analysis

In order to evaluate if the total phenolic contents (TPC) contributes to the antioxidant activity evaluated with DPPH and ORAC methodologies, a correlation analysis was carried out between TPC values with DPPH and ORAC results. Also, one-way analysis of variance (ANOVA) followed by Tukey's post hoc test was applied to TPC, DPPH and ORAC values, and differences were considered significant at $p < 0.05$.

4. Conclusions

The qualitative analysis of phenolic-enriched extracts of the only commercial cultivars of *M. domestica* (Anna cultivar) and *P. domestica* (Satsuma cultivar) in Costa Rica, using UPLC-DAD-ESI-MS techniques, shows 52 compounds characterized, distributed as 21 proanthocyanidins, including procyanidin A-type and B-type dimers and trimers, B-type tetramer and pentamers, flavan-3-ol monomers and gallates; 15 flavonoids, including kaempferol, quercetin and naringenin derivatives, and eight phenolic acid derivatives in both fruits; as well as chalcones and isoprenoid glycosides in Anna apples. These findings constitute the first report of such a high number and diversity of compounds in skins of one single plum cultivar and of the presence of proanthocyanidin pentamers in apple skins. Also, it is the first time that such a large number of glycosylated flavonoids and proanthocyanidins are reported in skins and flesh of a single plum cultivar. Further, significant negative correlation was found for both apple and plum samples between TPC and DPPH antioxidant values, especially for plum fruits ($R = -0.981$, $p < 0.05$) as well as significant positive correlation between TPC and ORAC, also especially for plum fruits ($R = 0.993$, $p < 0.05$). DPPH and ORAC methods show high values for all samples, especially for fruits skins, thus indicating the potential value of these extracts. The presence of procyanidin tetramers and pentamers in apple skin could be responsible for the higher antioxidant potential, in agreement with reports indicating higher antioxidant values related to the presence of this type of procyanidin oligomers [20,21]. Further purification or fractioning of these extracts would be important to evaluate their structure-bioactivity relationship and, for instance, specific proanthocyanidin structures effect on epithelial gastrointestinal cancer cells, of particular relevance due to promising related results [3,48] and the fact that proanthocyanidins low absorption make gut epithelial cells [49] likely one of the main tissues where these compounds can actually exert their biological effects.

Acknowledgments: This work was partially funded by a grant from FORINVES (Ref FV-0028-2013/115-B4-515). Authors also thank financial support from the University of Costa Rica and the Technological Institute of Costa Rica. Special thanks are due to Alonso Quesada from Costa Rican National Herbarium for his support with the vouchers.

Author Contributions: Mirtha Navarro participated in the conception and design of the study. Mirtha Navarro, Pei Chen, Silvia Quesada, Gabriela Azofeifa, Felipe Vargas and Diego Alvarado were involved in technical work and interpretation of data. Elizabeth Arnaez and Ileana Moreira participated in fruit collection, identification and initial samples treatment. Mirtha Navarro drafted the manuscript that was revised and approved by all the authors.

Conflicts of Interest: The authors declare no conflicts of interest.

References

1. Dai, J.; Mumper, R.J. Plant Phenolics: Extraction, Analysis and Their Antioxidant and Anticancer Properties. *Molecules* **2010**, *15*, 7313–7352. [CrossRef] [PubMed]
2. Quideau, S.; Deffieux, D.; Douat-Casassus, C.; Pouységu, L. Plant Polyphenols: Chemical Properties, Biological Activities, and Synthesis. *Angew. Chem. Int. Ed.* **2011**, *50*, 586–621. [CrossRef] [PubMed]
3. Eberhardt, M.; Lee, C.; Liu, R.H. Antioxidant activity of fresh apples. *Nature* **2000**, *405*, 903–904. [CrossRef] [PubMed]
4. Wolfe, K.; Wu, X.; Liu, R.H. Antioxidant activity of apple peels. *J. Agric. Food Chem.* **2003**, *51*, 609–614. [CrossRef] [PubMed]
5. Halvorsen, B.L.; Holte, K.; Myhrstad, M.C.W.; Barikmo, I.; Hvattum, E.; Remberg, S.F.; Wold, A.B.; Haffner, K.; Baugerod, H.; Andersen, L.F.; et al. A systematic screening of total antioxidant in dietary plants. *J. Nutr.* **2002**, *132*, 461–471. [CrossRef] [PubMed]
6. Noratto, G.; Porter, W.; Byrne, D.; Cisneros, L. Identifying Peach and Plum Polyphenols with Chemopreventive Potential against Estrogen-Independent Breast Cancer Cells. *J. Agric. Food Chem.* **2009**, *57*, 5219–5226. [CrossRef] [PubMed]
7. Hagen, S.F.; Borge, G.I.; Bengtsson, G.B.; Bilger, W.; Berge, A.; Haffner, K.; Solhaug, K.A. Phenolic contents and other health and sensory related properties of apple fruit (*Malus domestica* Borkh. Cv. Aroma): Effect of postharvest UV-B irradiation. *Postharvest Biol. Technol.* **2007**, *45*, 1–10. [CrossRef]

8. Khanizadeh, S.; Tsao, R.; Rekika, D.; Yang, R.; Charles, M.T.; Rupasinghe, V. Polyphenol composition and total antioxidant capacity of selected apple genotypes for processing. *J. Food Compost. Anal.* **2008**, *21*, 396–401. [CrossRef]
9. Tsao, R.; Yang, R.; Young, C.; Zhu, H. Polyphenolic Profiles in Eight Apple Cultivars Using High-Performance Liquid Chromatography (HPLC). *J. Agric. Food Chem.* **2003**, *57*, 6347–6353. [CrossRef] [PubMed]
10. Giomaro, G.; Karioti, A.; Bucchini, A.; Giamperi, L.; Ricci, D.; Fraternale, D. Polyphenols profile and antioxidant activity of skin and pulp of a rare apple from Marche region (Italy). *Chem. Cent. J.* **2014**, *8*, 45. [CrossRef] [PubMed]
11. Zupan, A.; Mikulic-Petkovsek, M.; Slantnar, A.; Stampar, F.; Veberic, R. Individual phenolic response and peroxidase activity in peel of differently sun-exposed apples in the period favorable for sunburn occurrence. *J. Plant Physiol.* **2014**, *171*, 1706–1712. [CrossRef] [PubMed]
12. Fang, N.; Yu, S.; Prior, R.L. LC/MS/MS Characterization of Phenolic Constituents in Dried Plums. *J. Agric. Food Chem.* **2002**, *50*, 3579–3585. [CrossRef] [PubMed]
13. Nakatani, N.; Kayano, S.; Kikuzaki, H.; Sumino, K.; Katagiri, K.; Mitani, T. Identification, quantitative determination, and antioxidative activities of chlorogenic acid isomers in prune (*Prunus domestica* L.). *J. Agric. Food Chem.* **2000**, *48*, 5512–5516. [CrossRef] [PubMed]
14. Slimstead, R.; Vangdal, E.; Brede, C. Analysis of Phenolic Compounds in Six Norwegian Plum Cultivars (*Prunus domestica* L.). *J. Agric. Food Chem.* **2009**, *57*, 11370–11375. [CrossRef] [PubMed]
15. Treutter, D.; Wang, D.; Farag, M.A.; Baires, G.D.; Rühmann, S.; Neumüller, M. Diversity of phenolic profiles in the fruit skin of Prunus domestica plums and related species. *J. Agric. Food Chem.* **2012**, *60*, 12011–12019. [CrossRef] [PubMed]
16. Jaiswal, R.; Karakose, H.; Ruhmann, S.; Goldner, K.; Neumuller, M.; Treutter, D.; Kuhnert, N. Identification of phenolic compounds in plum fruits (*Prunus salicina* L. and *Prunus domestica* L.) by HPLC/MS and characterization of varieties by quantitative phenolic fingerprints. *J. Agric. Food Chem.* **2013**, *61*, 12020–12031. [CrossRef] [PubMed]
17. Tomas-Barberan, F.; Gil, M.I.; Cremin, P.; Waterhouse, A.L.; Hess-Pierce, B.; Adel, A.; Kader, A.A. HPLC-DAD-ESIMS Analysis of Phenolic Compounds in Nectarines, Peaches, and Plums. *J. Agric. Food Chem.* **2001**, *49*, 4748–4760. [CrossRef] [PubMed]
18. Nunes, C.; Guyot, S.; Marnet, N.; Barros, A.S.; Saraiva, J.A.; Renard, C.M.; Coimbra, M.A. Characterization of plum procyanidins by thiolytic depolymerization. *J. Agric. Food Chem.* **2008**, *56*, 5188–5196. [CrossRef] [PubMed]
19. Navarro, M.; Lebron, R.; Quintanilla, J.E.; Cueva, C.; Hevia, D.; Quesada, S.; Gabriela Azofeifa, G.; Moreno, M.V.; Monagas, M.; Bartolome, B. Proanthocyanidin Characterization and Bioactivity of Extracts from Different Parts of *Uncaria tomentosa* L. (Cat's Claw). *Antioxidants* **2017**, *6*, 12. [CrossRef] [PubMed]
20. Es-Safi, N.E.; Guyot, S.; Ducrot, P.H. NMR, ESI/MS, and MALDI-TOF/MS analysis of pear juice polymeric proanthocyanidins with potent free radical scavenging activity. *J. Agric. Food Chem.* **2006**, *54*, 6969–6977. [CrossRef] [PubMed]
21. Spranger, I.; Sun, B.; Mateus, A.M.; Freitas, V.; Ricardo-da-Silva, J.M. Chemical characterization and antioxidant activities of oligomeric and polymeric procyanidin fractions from grape seeds. *Food Chem.* **2008**, *108*, 519–532. [CrossRef] [PubMed]
22. Arnous, A.; Anne, S.; Meyer, A.S. Comparison of methods for compositional characterization of grape (*Vitis vinifera* L.) and apple (Malus domestica) skins. *Food Bioprod. Process.* **2008**, *86*, 79–86. [CrossRef]
23. Lee, J.; Jeong, M.C.; Ku, K.H. Chemical, physical, and sensory properties of 1-MCP-treated Fuji apple (*Malus domestica* Borkh.) fruits after long-term cold storage. *Appl. Biol. Chem.* **2017**. [CrossRef]
24. Manzoor, M.; Anwar, F.; Saari, N.; Ashraf, M. Variations of Antioxidant Characteristics and Mineral Contents in Pulp and Peel of Different Apple (*Malus domestica* Borkh.) Cultivars from Pakistan. *Molecules* **2012**, *17*, 390–407. [CrossRef] [PubMed]
25. Zhang, T.; Wei, X.; Miao, Z.; Hassan, H.; Song, Y.; Fan, M. Screening for antioxidant and antibacterial activities of phenolics from *Golden Delicious* apple pomace. *Chem. Cent. J.* **2016**, *10*, 47. [CrossRef] [PubMed]
26. Cinquantaa, L.; Di Matteo, M.; Estia, M. Physical pre-treatment of plums (*Prunus domestica*). Part 2. Effect on the quality characteristics of different prune cultivars. *Food Chem.* **2002**, *79*, 233–238. [CrossRef]

27. Rop, O.; Jurikova, T.; Mlcek, J.; Kramarova, D.; Sengee, Z. Antioxidant activity and selected nutritional values of plums (*Prunus domestica* L.) typical of the White Carpathian Mountains. *Sci. Hortic.* **2009**, *122*, 545–549. [CrossRef]

28. Navarro, M.; Sanchez, F.; Murillo, R.; Martín, P.; Zamora, W.; Monagas, M.; Bartolomé, B. Phenolic assesment of *Uncaria tomentosa* L. (Cat's Claw): Leaves, stem, bark and wood extracts. *Molecules* **2015**, *20*, 22703–22717. [CrossRef] [PubMed]

29. Hamed, A.I.; Al-Ayed, A.S.; Moldoch, J.; Piacente, S.; Oleszek, W.; Stochmal, A. Profiles analysis of proanthocyanidins in the argun nut (*Medemia argun*—An ancient Egyptian palm) by LC-ESI-MS/MS. *J. Mass Spectrom.* **2014**, *49*, 306–315. [CrossRef] [PubMed]

30. Kammerer, D.; Claus, A.; Carle, R.; Schieber, A. Polyphenol Screening of Pomace from Red and White Grape Varieties (*Vitis vinifera* L.) by HPLC-DAD-MS/MS. *J. Agric. Food Chem.* **2004**, *52*, 4360–4367. [CrossRef] [PubMed]

31. Karonen, M.; Loponen, J.; Ossipov, V.; Pihlaja, K. Analysis of procyanidins in pine bark with reversed-phase and normal-phase high-performance liquid chromatography-electrospray ionization mass spectrometry. *Anal. Chim. Acta* **2004**, *522*, 105–112. [CrossRef]

32. Sarnoski, P.J.; Johnson, J.V.; Reed, K.A.; Tanko, J.M.; O'Keefe, S.F. Separation and characterisation of proanthocyanidins in Virginia type peanut skins by LC-MSn. *Food Chem.* **2012**, *131*, 927–939. [CrossRef]

33. Ojwang, L.O.; Yang, L.Y.; Dykes, L.; Awika, J. Proanthocyanidin profile of cowpea (*Vigna unguiculata*) reveals catechin-*O*-glucoside as the dominant compound. *Food Chem.* **2013**, *139*, 35–43. [CrossRef] [PubMed]

34. March, R.E.; Miao, X.S. A fragmentation study of kaempferol using electrospray quadrupole time-of-flight mass spectrometry at high mass resolution. *Int. J. Mass Spectrom.* **2004**, *231*, 157–167. [CrossRef]

35. Schültz, K.; Kammerer, D.; Carle, R.; Schieber, A. Identification and Quantification of Caffeoylquinic Acids and Flavonoids from Artichoke (*Cynara scolymus* L.) Heads, Juice, and Pomace by HPLC-DAD-ESI/MSn. *J. Agric. Food Chem.* **2004**, *52*, 4090–4096. [CrossRef] [PubMed]

36. Lin, L.Z.; Harnly, J.M. A screening method for the identification of glycosylated flavonoids and other phenolic compounds using a standard analytical approach for all plant materials. *J. Agric. Food Chem.* **2007**, *55*, 1084–1096. [CrossRef] [PubMed]

37. Gruz, J.; Novák, O.; Strnad, M. Rapid analysis of phenolic acids in beverages by UPLC–MS/MS. *Food Chem.* **2008**, *111*, 789–794. [CrossRef]

38. Bylund, D.; Norström, S.H.; Essén, S.A.; Lundström, U.S. Analysis of low molecular mass organic acids in natural waters by ion exclusion chromatography tandem mass spectrometry. *J. Chromatogr. A* **2007**, *1176*, 89–93. [CrossRef] [PubMed]

39. Shao, X.; Bai, N.; He, K.; Ho, C.-T.; Yang, C.S.; Sang, S. Apple Polyphenols, Phloretin and Phloridzin: New Trapping Agents of Reactive Dicarbonyl Species. *Chem. Res. Toxicol.* **2008**, *21*, 2042–2050. [CrossRef] [PubMed]

40. Schwab, W.; Schreier, P. Vomifoliol 1-0-β-d-xylopyranosyl-6-o-β-d-glucopyranoside: A disaccharide glycoside from apple fruit. *Phytochemistry* **2008**, *29*, 161–164. [CrossRef]

41. Vineetha, V.P.; Girija, S.; Soumya, R.S.; Raghu, K.G. Polyphenol-rich apple (*Malus domestica* L.) peel extract attenuates arsenic trioxide induced cardiotoxicity in H9c2 cells via its antioxidant activity. *Food Funct.* **2014**, *5*, 502–511. [CrossRef] [PubMed]

42. Shimamura, T.; Sumikura, Y.; Yamazaki, T.; Tada, A.; Kashiwagi, T.; Ishikawa, H.; Matsui, T.; Sugimoto, N.; Akiyama, H.; Ukeda, H. Applicability of the DPPH assay for evaluating the antioxidant capacity of food additives—Inter-laboratory evaluation study. *Anal. Sci.* **2014**, *30*, 717–721. [CrossRef] [PubMed]

43. Wong, S.P.; Leong, L.P.; Koh, J.H. Antioxidant activities of aqueous extracts of selected plants. *Food Chem.* **2006**, *99*, 775–783. [CrossRef]

44. Lizcano, L.J.; Bakkali, F.; Ruiz-Larrea, B.; Ruiz-Sanz, J.I. Antioxidant activity and polyphenol content of aqueous extracts from Colombia Amazonian plants with medicinal use. *Food Chem.* **2010**, *119*, 1566–1570. [CrossRef]

45. Singleton, V.; Rossi, J. Colorimetry of total phenolics with phosphomolybdic-phosphotungstic acid reagents. *Am. J. Enol. Vitic.* **1965**, *16*, 144–158.

46. Navarro, M.; Moreira, I.; Arnaez, E.; Quesada, S.; Azofeifa, G.; Alvarado, D.; Monagas, M.J. Proanthocyanidin Characterization, Antioxidant and Cytotoxic Activities of Three Plants Commonly Used in Traditional Medicine in Costa Rica: *Petiveria alliaceae* L., *Phyllanthus niruri* L. and *Senna reticulata* Willd. *Plants* **2017**, *6*, 50. [CrossRef] [PubMed]

47. Davalos, A.; Gomez-Cordoves, C.; Bartolome, B. Extending applicability of the oxygen radical absorbance capacity (ORAC-Fluorescein) assay. *J. Agric. Food Chem.* **2004**, *52*, 48–54. [CrossRef] [PubMed]

48. Navarro, M.; Zamora, W.; Quesada, S.; Azofeifa, G.; Alvarado, D.; Monagas, M. Fractioning of Proanthocyanidins of *Uncaria tomentosa*. Composition and Structure-Bioactivity Relationship. *Antioxidants* **2017**, *6*, 60. [CrossRef] [PubMed]

49. Monagas, M.; Urpi-Sarda, M.; Sanchez-Patán, F.; Llorach, R.; Garrido, I.; Gómez-Cordoves, C.; Andres-Lacueva, C.; Bartolome, B. Insights into the metabolism and microbial biotransformation of dietary flavan-3-ols and the bioactivity of their metabolites. *Food Funct.* **2010**, *1*, 233–253. [CrossRef] [PubMed]

Article

Detection of Lard in Cocoa Butter—Its Fatty Acid Composition, Triacylglycerol Profiles, and Thermal Characteristics

Marliana Azir [1], Sahar Abbasiliasi [2,3,*], Tengku Azmi Tengku Ibrahim [4,5], Yanty Noorzianna Abdul Manaf [1], Awis Qurni Sazili [1,6,7] and Shuhaimi Mustafa [1,2,3,*]

[1]   Halal Products Research Institute, Universiti Putra Malaysia, 43400 UPM Serdang, Selangor, Malaysia; marlianaazir@gmail.com (M.A.); yanty@upm.edu.my (Y.N.A.M.); awis@upm.edu.my (A.Q.S.)

[2]   Department of Microbiology, Faculty of Biotechnology and Biomolecular Sciences, Universiti Putra Malaysia, 43400 UPM Serdang, Selangor, Malaysia

[3]   Bioprocessing and Biomanufacturing Research Center, Universiti Putra Malaysia, 43400 UPM Serdang, Selangor, Malaysia

[4]   Institute of Bioscience, Universiti Putra Malaysia, 43300 Serdang, Selangor, Malaysia; tengkuazmi@upm.edu.my

[5]   Faculty of Veterinary Medicine, Universiti Putra Malaysia, 43400 UPM Serdang, Selangor, Malaysia

[6]   Department of Animal Science, Faculty of Agriculture, Universiti Putra Malaysia, 43400 UPM Serdang, Selangor, Malaysia

[7]   Institute of Tropical Agriculture and Food Security, Universiti Putra Malaysia, 43400 UPM Serdang, Selangor, Malaysia

*   Correspondences: sahar@upm.edu.my (S.A.); shuhaimi@upm.edu.my (S.M.); Tel.: +60-3-8941-2624 (S.A. & S.M.)

Received: 25 September 2017; Accepted: 16 October 2017; Published: 9 November 2017

Abstract: The present study investigates the detection of lard in cocoa butter through changes in fatty acids composition, triacylglycerols profile, and thermal characteristics. Cocoa butter was mixed with 1% to 30% (*v/v*) of lard and analyzed using a gas chromatography flame ionization detector, high performance liquid chromatography, and differential scanning calorimetry. The results revealed that the mixing of lard in cocoa butter showed an increased amount of oleic acid in the cocoa butter while there was a decrease in the amount of palmitic acid and stearic acids. The amount of POS, SOS, and POP also decreased with the addition of lard. A heating thermogram from the DSC analysis showed that as the concentration of lard increased from 3% to 30%, two minor peaks at −26 °C and 34.5 °C started to appear and a minor peak at 34.5 °C gradually overlapped with the neighbouring major peak. A cooling thermogram of the above adulterated cocoa butter showed a minor peak shift to a lower temperature of −36 °C to −41.5 °C. Values from this study could be used as a basis for the identification of lard from other fats in the food authentication process.

Keywords: cocoa butter; lard; fatty acids methyl ester; triacylglycerol; adulteration; food

---

1. Introduction

Worldwide, there is an ever increasing demand by consumers for information and confidence pertaining to the origin and content of purchased food. In this respect, food manufacturers have no alternative but to provide and confirm the authenticity of the origin of their food ingredients. This pressing demand, accompanied by legislative and regulatory drives, has increased the complexity and level of regulation imposed on food production. Protecting consumer rights becomes the foremost issue and constitutes important challenges facing the food industry [1].

The replacement of expensive oils by comparatively cheaper ones is a common practice from an economic point of view. There is also a tendency for oil to be substituted in view of its high price, increased demand, limited availability, and accessibility. Cocoa butter is a byproduct of cocoa. As it constitute an expensive component of chocolate and plays an important role in the melting properties of chocolate, its availability in the market is most often unpredictable. Its demand in the food and pharmaceutical industries is very high by virtue of its physical properties and organoleptic qualities. For a long time, there has been considerable effort to replace it either fully or partially with other vegetable fats, the so-called cocoa butter alternatives, which would be much cheaper. Lard could be another alternative for cocoa butter as it is the cheapest form of fat readily available for use by the food industries. Lard or industrially modified lard could be effectively blended with other vegetable oils to produce shortenings, margarines, and other food oils. There is, however, a limitation on the use of this animal product in the food industry from the perspective of the Muslim religion, in addition to the risk of biological complications and health risks associated with daily intake [2].

The ever increasing price of organic products and limited availability brought about by high demand has led to an increased number of fraudulent practices in the food industry which calls for a reexamination of the procedures with respect to their authentication to reassure both consumers' confidence and fair trade practices. In this context, it is pertinent to develop analytical procedures capable of detecting fraudulent practices and protect the consumers from misleading labeling and unsubstantiated claims.

Using physical properties such as the refractive index, viscosity, melting point, saponification, and iodine value are no longer practical to detect adulteration in view of the availability of more current, sophisticated procedures and approaches. However, each oil and fat has a specific component at a known level and their presence and quantity should be considered as a detection tool. Therefore, advanced and sophisticated methods with high sensitivity to detect and quantify adulteration need to be given due consideration [3].

Several methods have been employed to detect lard in foods and food products which include DNA-based polymerase chain reaction (PCR), Fourier transform infrared spectroscopy (FTIR), electronic nose technology (e-nose), differential scanning calorimetry (DSC), and chromatographic-based techniques (Table 1). However, to date there is no report published on the application of a gas chromatography flame ionization detector (GC-FID), high performance liquid chromatography (HPLC), and DSC for the detection and quantification of lard in cocoa butter. Therefore, this study is aimed at determining the level of lard adulteration in cocoa butter using DSC, HPLC, and GC-FID.

Table 1. Methods for the detection of lard in foods and food products.

Issues in Food Sample	Method of Detection	References
DNA-based PCR method		
Pork and lard in food products	cyt *b* PCR-RFLP	[4]
Lard in food products (sausages and casings, bread and biscuits)	cyt *b* PCR-RFLP	[5]
Lard detection in chocolate	Porcine-specific real-time PCR	[6]
Fourier transform infrared spectroscopy		
Lard mixed with other animal fats	FTIR with PLS	[7]
Lard mixed with animal fats	FTIR with PLS	[8]
Lard in cake formulation	FTIR with PLS	[9]
Lard in chocolate and chocolate products	FTIR with PLS	[2]
Lard in biscuit	FTIR with PLS	[10]
Lard mixed with lamb, cow and chicken body fats	FTIR with PLS and DA	[11]
Lard mixed with cod liver oil	FTIR with PLS and DA	[12]
Lard in other animal fats	FTIR with PLS and DA	[11]
Lard in virgin coconut oil (VCO)	FTIR with PLS and DA	[13]
Lard in vegetable oils	FTIR with PLS, PCR and DA	[14]
Lard in edible fats and oil	FTIR with PCA and CA	[15]
Lard in cream cosmetics	FTIR with PLS and PCR	[16]
Lard in frying oil	FTIR with PLS and DA	[17]
Lard in chocolate	FTIR with PLS and PCA	[18]
Lard in ink extracted from printed food packaging	FTIR with PCA and SIMCA	[19]
Electronic nose technology		
Lard in edible oil	E-nose	[20]

Table 1. *Cont.*

Issues in Food Sample	Method of Detection	References
Differential scanning calorimetry		
Lard and randomized lard in RBD palm oil	DSC	[21]
Monitoring lard in canola oil	DSC	[22]
Lard adulteration	DSC	[22]
Lard in cooking oil	DSC	[23]
Lard in sunflower oil	DSC	[24]
Lard in canola oil	DSC	[25]
Lard in virgin coconut oil	DSC	[13]
Lard in butter	DSC	[26]
Chromatographic-based techniques		
Lard in meat products	HPLC	[27]
Lard in meat lipids	HPLC	[28]
Lard in animal fats and vegetable oils	HPLC	[28]
Lard in fried oils	HPLC	[29]
Lard in meat lipids	GLC (FID detector)	[30]
Lard in milk lipids	GLC (FID detector)	[31]
Lard in milk lipids	GLC (FID detector)	[32]
Lard in fried oils	GLC (FID detector)	[29]
Lard in animal fats	LC-MS	[33]
Lard in animal fats	GC-FID	[31]
Lard in vegetable oils	GC-FID	[28]
Lard in milk fat	GC	[34]
Lard in vegetable oils	GC (FID detector)	[28]
Lard in animal fats	GC×GC–TOF-MS	[35]
Lard in animal fats	GC×GC–TOF-MS	[36]

PCA: principal component analysis; DA: discriminant analysis; CA: cluster analysis; PCR: principle component regression; PLS: partial least square; SIMCA: soft independent modeling class analogy; RBD: refined, bleached, deodorized; PCR-RFLP: polymerase chain reaction-restriction fragment length polymorphism; GLC: gas liquid chromatography; FID: flame ionization detector; LC-MS: liquid chromatography–mass spectrometry; GC: gas chromatography; TOF: time-of-flight; MS: mass spectrometry.

2. Materials and Methods

2.1. Sample Preparation and Supplies

Pig adipose tissue was obtained from a local market at Sri Serdang, Selangor, Malaysia, while cocoa butter (CB) was kindly donated by the Malaysian Palm Oil Board (MPOB). Acetone (C_3H_6O, $\geq$99.9%), acetonitrile (CH_3CN, $\geq$99.0%), chloroform ($CHCl_3$, $\geq$99.8%), anhydrous sodium sulfate (Na_2SO_4), cyclohexane (C_6H_{12}), Wij'sreagent (Iodine trichloride solution), potassium iodide (KI), sodium thiosulphate pentahydrate ($Na_2S_2O_3 \cdot 5H_2O$), and acetic acid glacial (CH_3CO_2H, $\geq$99.0%) were supplied from Orec, New Zealand. Triacylglycerol (TAG) standards and the sodium methoxide solution (CH_3ONa, 25 wt % in methanol) were sourced from Sigma-Aldrich (St. Louis, MO, USA). All chemicals used in this study were of analytical grade.

2.2. Extraction of Oil from Lard

The oil extraction procedure is as described by Hoffmann [37]. Briefly, the animal fat samples were diced into pieces measuring 0.5 cm × 0.5 cm, heated at 90–100 °C for 2 h, and strained through a triple folded muslin cloth to remove impurities. The melted fats were filtered through Whatman No. 2 filter paper containing Na_2SO_4 flushed with nitrogen to prevent oxidation [38] and stored in a tightly closed container at 4 °C until use.

2.3. Blend Preparation of Adulterated Cocoa Butter

Cocoa butter and lard were melted at 60 °C and blended in proportions of 99:1, 97:3, 95:5, 90:10, 85:15, 80:20, 75:25, and 70:30 (g/100 g).

2.4. Determination of Triacylglycerol Composition of Lard, Cocoa Butter, and Their Admixture

The determination of the TAG composition of lard, cocoa butter, and adulterated cocoa butter with 1% to 30% of lard was carried out according to the procedure described by Haryati et al. [39]. Briefly, 0.1 g of the sample was dissolved in 1 mL of chloroform. The composition of TAG was determined by HPLC (Waters Model 510, Waters Associates, Milford, MA, USA). The HPLC system was equipped with a LiChrosphere® RP-18 column (5 μm particle size, 12.5 cm × 34 mm) (Merck, Darmstadt, Germany) and RID detector (Model 410, Waters Associates, Milford, MA, USA). A mixture of acetone: acetonitrile (63.5:36.5) was used as the solvent for elution at a flow rate of 1.5 mL/min. The column temperature was maintained at 30 °C. All measurements were carried out in triplicate. Peaks for the respective samples were identified using a set of TAG standards.

2.5. Determination of Fatty Acid Composition of Lard, Cocoa Butter, and Their Admixture

Fatty acids methyl ester (FAME) was prepared according to the method of Marina et al. [40]. A total of 50 mg of each sample was dissolved in 0.8 mL of hexane and 0.2 mL of 1 M sodium methoxide. The mixture was vortexed for 1 min and 1 μL of the clear supernatant was subsequently injected into a gas chromatograph (Shimadzu GC-14 A) equipped with an FID detector (Shimadzu, Vienna, Austria). A polar capillary column BPX70 (0.32 mm internal diameter, 30 m length and 0.25 mm film thickness; SGE International Pty, Ltd., Victoria, Australia) was used at a column pressure of 10 psi. The initial column oven temperature was set at 90 °C, and programmed to increase to 220 °C at 15 °C/min (for 5 min), 2 °C/min (for 20 min), and 15 °C/min (for 1 min). Temperatures of the injector and detector were maintained at 240 °C. Peaks of the respective samples were identified by comparing their retention time with certified reference standards of FAME (Supelco, Bellefonte, PA, USA). The area percent of each fatty acid was calculated by dividing its peak area by the total peak area of the fatty acids identified.

2.6. Thermal Analysis of Lard, Cocoa Butter, and Their Admixture Using Differential Scanning Calorimetry

The thermal characteristics of the oils were analyzed using DSC (Mettler Toledo, Greifensee, Switzerland). Four to eight mg of oils was weighed into aluminium pans sealed hermetically and analyzed using a DSC Q100 instrument (TA Instruments, New Castle, DE, USA). Nitrogen (99.99% purity) was purged at a flow of 20 mL/min. The instrument was calibrated using indium (m.p. 156.6 °C, ΔH_f = 28.45 J/g) and n-dodecane (m.p. −9.65 °C, ΔH_f = 216.73 J/g), and an empty pan was used as reference [41]. All samples were subjected to the following temperature programs where the samples were cooled from 50 °C to −70 °C, held for 5 min, and heated from −70 °C to 50 °C at a rate of 5 °C/min. The melting and crystallisation parameters of each sample were obtained using Mettler Toledo STAR^e software system (STAR^e SW 9.20 software, Greifensee, Switzerland). Thermograms were analyzed with universal analysis software (Version 3.9A, TA Instruments, New Castle, DE, USA) to obtain the enthalpy (ΔH, J/g), T_{on} (°C), and T_{off} (°C) of the transitions (intersection of baseline and tangent at the transition) and peak temperature (Tp °C). The range of the transitions was calculated as the temperature difference between T_{on} and T_{off}. All experiments were determined in triplicate and the average of three measurements was used for data analysis.

2.7. Statistical Analysis

Data obtained were analyzed using a one-way analysis of variance (ANOVA) test. The Tukey's honest significant difference test (HSD) method was chosen for the post-hoc tests for each dataset. The Tukey Honestly Significant Difference (HSD) function of SPSS (Version 14.0, SPSS Inc., Chicago, IL, USA) was used to compare the means of all data. All statistics were based on a confidence level of 95%, and $p < 0.05$ was considered statistically significant.

3. Results and Discussion

3.1. Triacylglycerol Composition of Lard, Cocoa Butter, and Their Admixture

The TAG profiles of lard, cocoa butter, and their admixture are shown in Table 2. The distinction in the nature of TAG is the principal factor that makes fats different from one another and these variations affect the TAG separation. The total di-unsaturated TAGs in LD (48.03%) were higher compared to those of cocoa butter, while cocoa butter had higher total di-saturated TAG values (90.55%). Moreover, a relatively high proportion of L, Ln, and O present in lard was reflected by the shorter retention times of TAGs such as linolenoyldilinoleoylglycerol (LLLn), trilinolein (LLL), dilinoleoyloleoylglycerol (OLL), linoleoyldioleoylglycerol (OOL), dilinoleoylpalmitoylglycerol (PLL), linoleoyloleoylpalmitoylglycerol (POL), and dioleoypalmitoylglycerol (POO) (Figure 1). POL (20.21%), POS (18.58%), and POO (17.25%) in lard and POS (41.67%), SOS (28.47%), and POP (19.13%) in cocoa butter are the most abundant TAG. However, this total amount in cocoa butter is slightly different from that reported by Shukla [42], although in this research, POS (38.5%), SOS (30.30%), and POP (15.20%) were the major TAGs. This variation could be due to the different methods of extraction, ripeness values, cultivar types, growing conditions, and the origin of cocoa.

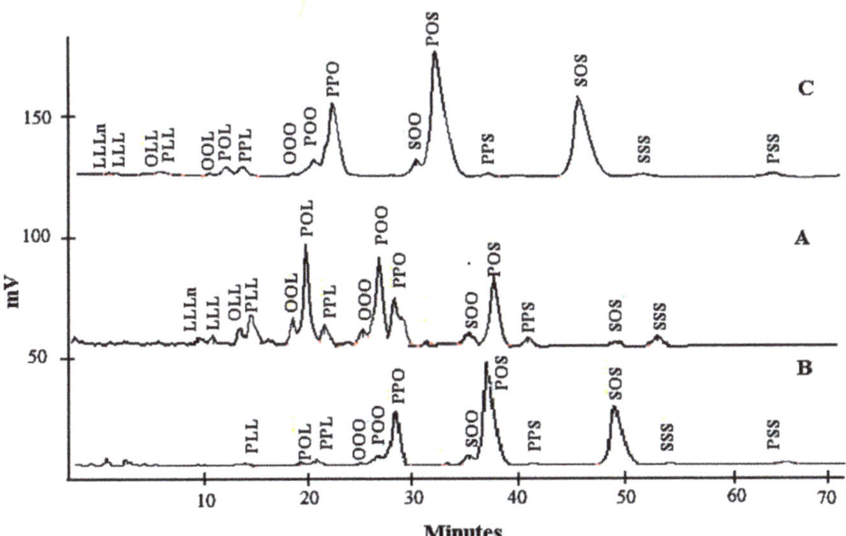

Figure 1. The chromatogram of lard (**A**), cocoa butter (**B**), and their admixture (**C**).

The levels of TAGs containing S viz. SOO, POS, PSS, and SOS were significantly different ($p < 0.05$) in cocoa butter and lard. The TAG profiles of fats resulted in a high proportion of unsaturated FAs for lard (55.06%) and cocoa butter (33.74%). Table 2 showed the TAG profiles when lard was added to cocoa butter from 1% to 30%. An adulterated sample with lard has additional TAGs viz. LLLn, LLL, OLL, OOL, PLL, OLL, POL, and POO. Cocoa butter adulterated with lard caused a slight increase in oleic-acid-predominating TAGs, while the palmitic acid containing TAGs decreased slightly. Kallio et al. [43] reported that lard has major saturated FA, especially palmitic acid, at the *sn*-2 position, which made it distinct from other fats and oils.

Table 2. TAG composition of lard, cocoa butter, and their admixtures.

	TAGs	Lard Concentration (%)									
		0 (CB)	1	3	5	10	15	20	25	30	100 (LD)
Unsaturated											
Tri-unsaturated	LLLn	nd < 0.04	0.04 (0.00) [f]	0.05 (0.00) [e,f]	0.05 (0.00) [e,f]	0.06 (0.00) [e]	0.07 (0.00) [d]	0.11 (0.00) [c]	0.35 (0.00) [b]	0.36 (0.01) [a]	0.88 (0.00) [g]
	LLL	nd < 0.04	0.05 (0.00) [h]	0.06 (0.00) [g]	0.10 (0.00) [f]	0.11 (0.00) [e]	0.13 (0.00) [d]	0.13 (0.00) [c]	0.15 (0.00) [b]	0.17 (0.00) [a]	1.24 (0.01) [h]
	OLL	nd < 0.04	0.05 (0.00) [h]	0.08 (0.00) [g]	0.15 (0.00) [f]	0.25 (0.00) [e]	0.35 (0.00) [d]	0.51 (0.00) [c]	0.52 (0.00) [b]	0.85 (0.01) [a]	2.94 (0.01) [i]
	OOL	nd < 0.04	0.19 (0.00) [h]	0.21 (0.01) [g]	0.37 (0.00) [f]	0.45 (0.00) [e]	0.58 (0.00) [d]	0.87 (0.00) [c]	1.17 (0.00) [b]	1.29 (0.00) [a]	4.38 (0.01) [i]
	OOO	0.24 (0.00) [h]	0.30 (0.00) [g]	0.32 (0.00) [g]	0.35 (0.01) [f]	0.42 (0.00) [e]	0.58 (0.00) [d]	0.73 (0.00) [c]	0.93 (0.00) [b]	0.98 (0.00) [a]	2.28 (0.01) [i]
	Sub total	0.24 (0.11)	0.63 (0.12)	0.72 (0.12)	1.02 (0.15)	1.29 (0.18)	1.71 (0.24)	2.35 (0.34)	3.12 (0.42)	3.65 (0.46)	11.72 (1.40)
Di-unsaturated	PLL	0.22 (0.01) [g]	0.10 (0.00) [i]	0.16 (0.00) [h]	0.35 (0.00) [f]	0.69 (0.00) [e]	1.01 (0.00) [d]	1.68 (0.01) [c]	2.19 (0.01) [b]	2.57 (0.02) [a]	7.36 (0.08) [j]
	POL	0.66 (0.00) [i]	1.28 (0.00) [h]	1.49 (0.00) [g]	1.69 (0.00) [f]	2.39 (0.01) [e]	4.00 (0.00) [d]	4.29 (0.01) [c]	5.68 (0.01) [b]	6.18 (0.00) [a]	20.21 (0.01) [j]
	POO	3.12 (0.01) [i]	3.30 (0.00) [h]	3.43 (0.01) [g]	3.65 (0.01) [f]	4.08 (0.01) [e]	4.76 (0.01) [d]	5.75 (0.01) [c]	6.98 (0.01) [b]	7.13 (0.03) [a]	17.25 (0.01) [j]
	SOO	3.46 (0.05) [a]	3.40 (0.00) [a]	3.32 (0.00) [b]	3.21 (0.01) [c]	3.19 (0.01) [c]	3.14 (0.00) [c,d]	3.11 (0.00) [d,e]	3.09 (0.01) [d,e]	3.05 (0.03) [e]	3.21 (0.00) [c]
	Sub total	7.46 (1.66)	8.08 (1.61)	8.40 (1.57)	8.90 (1.51)	10.35 (1.44)	12.91 (1.62)	14.83 (1.73)	17.94 (2.23)	18.93 (2.26)	48.03 (8.04)
	Total unsaturated	**7.70 (5.11)**	**8.71 (5.27)**	**9.12 (5.43)**	**9.92 (5.57)**	**11.64 (6.41)**	**14.62 (7.92)**	**17.18 (8.82)**	**21.06 (10.48)**	**22.58 (10.80)**	**59.75 (25.68)**
Saturated											
Di-saturated	POP	19.13 (0.02) [a]	18.77 (0.01) [b]	18.64 (0.00) [c]	18.56 (0.01) [d]	18.47 (0.01) [e]	18.31 (0.01) [f]	18.01 (0.00) [g]	17.64 (0.01) [h]	17.29 (0.01) [i]	3.21 (0.00) [j]
	PPL	1.28 (0.00) [g]	1.76 (0.00) [f]	1.80 (0.01) [f]	1.81 (0.01) [f]	1.93 (0.02) [e]	2.02 (0.00) [d]	2.13 (0.04) [c]	2.30 (0.00) [b]	2.39 (0.01) [a]	4.35 (0.00) [h]
	POS	41.67 (0.01) [a]	41.21 (0.01) [b]	40.90 (0.01) [c]	40.32 (0.01) [d]	39.41 (0.01) [e]	37.79 (0.01) [f]	36.57 (0.04) [g]	34.91 (0.01) [h]	34.01 (0.01) [i]	18.58 (0.01) [j]
	SOS	28.47 (0.02) [a]	27.67 (0.01) [b]	27.55 (0.01) [c]	27.37 (0.01) [d]	26.51 (0.00) [e]	25.15 (0.01) [f]	23.87 (0.01) [g]	21.71 (0.02) [h]	20.96 (0.01) [i]	1.32 (0.00) [j]
	Sub total	90.55 (16.98)	89.41 (16.54)	88.89 (16.40)	88.06 (16.17)	86.32 (15.68)	83.27 (14.90)	80.58 (14.29)	76.56 (13.43)	74.65 (13.01)	27.46 (7.91)
Tri-saturated	PPS	0.27 (0.00) [i]	0.32 (0.00) [h]	0.40 (0.00) [g]	0.46 (0.00) [f]	0.51 (0.00) [e]	0.58 (0.00) [d]	0.68 (0.00) [c]	0.74 (0.00) [b]	1.09 (0.00) [a]	1.99 (0.00) [j]
	SSS	0.31 (0.00) [i]	0.41 (0.01) [h]	0.48 (0.00) [g]	0.50 (0.00) [f]	0.52 (0.00) [e]	0.59 (0.00) [d]	0.67 (0.00) [c]	0.87 (0.00) [b]	0.99 (0.00) [a]	2.89 (0.01) [j]
	PSS	1.17 (0.01) [a]	1.15 (0.00) [b]	1.11 (0.01) [c]	1.06 (0.00) [d]	1.01 (0.00) [e]	0.93 (0.00) [f]	0.90 (0.00) [g]	0.79 (0.00) [h]	0.69 (0.00) [i]	nd <0.04
	Sub total	1.75 (0.51)	1.88 (0.46)	1.99 (0.39)	2.02 (0.34)	2.04 (0.29)	2.10 (0.20)	2.25 (0.13)	2.40 (0.07)	2.77 (0.21)	4.88 (1.46)
	Total saturated	**92.30 (62.79)**	**91.29 (61.89)**	**90.88 (61.45)**	**90.08 (60.84)**	**88.36 (59.59)**	**88.36 (59.59)**	**82.83 (55.39)**	**78.96 (52.44)**	**77.42 (50.83)**	**32.34 (15.97)**

Each value represents the mean ± SD of triplicate analyses; Means within the same row with different superscripts are significantly different ($p < 0.05$); Abbreviations: TAG, triacylglycerol; CB, cocoa butter; LD, lard; P, palmitic; O, oleic; L, linoleic; S, stearic; nd, not detected.

3.2. Fatty Acid Methyl Ester Composition of Lard, Cocoa Butter, and Their Admixtures

Fatty acids (FAs) are essential components of edible fats and oils in which they can be found in the ester form with a glycerol backbone (triglycerides). Besides, FA compositions differ from one source to another. Therefore, FA profiles can be used for determining the purity or authenticity of animal fats. The quantitative analysis of FA composition is essential in food research with regards to the nutritional value content. In this study, the FAME compositions of lard, cocoa butter, and the mixture of lard in cocoa butter from 1% to 30% determined using GC-FID are presented in Table 3.

Three major fatty acids of lard (C18:0, C18:1, C18:2, and C16:0) in this study (0.36, 19.29, 32.41, 22.55) were lower compared to the values of 11.53, 24.64, and 17.29 reported by Nizar et al. [44]. However, the results in the present study concurred with those reported by Cheong et al. [45] and Nurjuliana et al. [46]. Edwards et al. [47] reported that fatty acid composition is influenced by the species, sex, and diet of animals.

Lard could be differentiated from cocoa butter having C10:0 (0.17%), C12:0 (1.44%), C15:0 (0.09%), C16:1 (1.22%), C17:0 (0.58%), and C18:3 (1.09%). Lard also has a higher total unsaturated fatty acid percentage (53.86%) compared to cocoa butter (33.74%). However, lard shared similar characteristics with that of cocoa butter having C16:0, C18:0, and C20:0, resulting in differences in their SFA contents. Cocoa butter has a total saturated fatty acid value of 66.27%.

Hence, from the above-mentioned findings, due to the addition of lard to cocoa butter from 1% to 30%, the amount of palmitic acid (C16:0), stearic acid (C18:0), capric acid (C10:0), lauric acid (C12:0), pentadecyclic acid (C15:0), palmitoleic acid (C16:1), margaric acid (C17:0), and linolenic acid (C18:3) increased. Increasing the proportion of lard in cocoa butter decreased the SFA from 65.66% to 59.23% and increased the USFA from 34.35% to 40.79%.

Table 3. Composition of fatty acids in lard, cocoa butter, and their admixtures.

FAs	Lard Concentration (%)									
	0 (CB)	1	3	5	10	15	20	25	30	100 (LD)
C10:0	0	0	0	0	0.01 (0.01) [c]	0.02 (0.03) [b,c]	0.06 (0.00) [a,b]	0.06 (0.00) [a,b]	0.08 (0.01) [a]	0.17 (0.00) [d]
C12:0	0	0.12 (0.00) [e]	0.14 (0.01) [d,e]	0.16 (0.00) [d,e]	0.21 (0.00) [c,d]	0.28 (0.01) [c]	0.36 (0.06) [b]	0.40 (0.00) [a,b]	0.47 (0.01) [a]	1.44 (0.00) [f]
C15:0	0	0	0	0	0	0	0	0	0	0.09 (0.00) [b]
C16:0	27.27 (0.07) [a]	26.63 (0.08) [b]	26.40 (0.00) [c]	26.32 (0.02) [c]	26.11 (0.07) [d]	25.98 (0.04) [d]	25.55 (0.03) [e]	25.43 (0.02) [e]	24.75 (0.04) [f]	22.55 (0.00) [g]
C16:1	0	0.25 (0.00) [e]	0.26 (0.00) [e]	0.28 (0.01) [d,e]	0.30 (0.00) [d]	0.36 (0.02) [c]	0.40 (0.01) [b]	0.42 (0.01) [a,b]	0.44 (0.01) [a]	1.22 (0.00) [f]
C17:0	0	0.24 (0.00) [e]	0.25 (0.01) [e]	0.25 (0.00) [d,e]	0.26 (0.00) [d]	0.26 (0.00) [d]	0.29 (0.00) [c]	0.31 (0.01) [b]	0.32 (0.00) [a]	0.58 (0.00) [f]
C18:0	37.75 (0.05) [a]	37.45 (0.28) [a,b]	37.09 (0.01) [b,c]	36.74 (0.01) [c]	36.03 (0.01) [d]	35.24 (0.13) [e]	34.10 (0.04) [f]	33.66 (0.01) [g]	32.75 (0.00) [h]	0.36 (0.00) [i]
C18:1	30.92 (0.06) [e]	31.03 (0.21) [e]	31.37 (0.00) [d]	31.51 (0.00) [c,d]	31.73 (0.07) [b,c]	31.86 (0.00) [a,b]	31.87 (0.01) [a,b]	31.92 (0.02) [a,b]	32.13 (0.01) [a]	19.29 (0.09) [f]
C18:2	2.82 (0.03) [h]	2.88 (0.00) [h]	3.10 (0.00) [g]	3.36 (0.00) [f]	3.96 (0.00) [e]	4.67 (0.03) [d]	6.04 (0.04) [c]	6.48 (0.00) [b]	7.83 (0.00) [a]	32.41 (0.08) [i]
C18:3	0	0.19 (0.00) [e]	0.20 (0.00) [d,e]	0.20 (0.00) [d,e]	0.24 (0.00) [c,d]	0.26 (0.01) [c]	0.32 (0.03) [b]	0.35 (0.00) [a,b]	0.39 (0.01) [a]	1.09 (0.00) [f]
C20:0	1.25 (0.01) [a]	1.22 (0.01) [a,b]	1.21 (0.01) [a,b]	1.19 (0.00) [a,b]	1.16 (0.00) [b]	1.09 (0.04) [c]	1.03 (0.01) [d]	1.00 (0.00) [d]	0.86 (0.00) [e]	0.82 (0.00) [e]
Total SFA	66.27 (18.79)	65.66 (17.68)	65.09 (17.51)	64.66 (17.38)	63.78 (16.15)	62.87 (15.87)	61.39 (15.42)	60.86 (15.26)	59.23 (14.84)	44.94 (9.96)
Total USFA	33.74 (19.87)	34.35 (15.01)	34.93 (15.15)	35.35 (15.19)	36.23 (15.21)	37.15 (15.19)	38.63 (15.05)	39.17 (15.03)	40.79 (15.03)	53.86 (15.23)

Data are presented as means ± SD from triplicate determination. Means within the same row with different superscripts are significantly different ($p < 0.05$). Abbreviations: FAs, fatty acids; CB, cocoa butter; LD, lard; SFA, saturated fatty acids ; USFA, unsaturated fatty acids.

3.3. Thermal Analysis of Lard, Cocoa Butter, and Their Admixtures during Heating and Cooling Temperatures

The DSC representative heating thermograms of lard, cocoa butter, and their admixtures in a range of 1% to 30% are as shown in Figure 2a. Lard and cocoa butter were totally different in their heating profiles. Lard (J) had two major peaks which appeared at −4.00 °C and 28.78 °C and a small shoulder peak at 34.09 °C. The heating thermal properties were acquired at the onset of the exothermal reaction, at the offset of the endothermal reaction, the peaks, and enthalpy. There is a significant difference ($p < 0.05$) between the heating enthalpy of lard (29.86 J/g) and cocoa butter (86.81 J/g). Cocoa butter (A) exhibited the highest T_{off} (23.38 °C) which could be due to a larger, highly saturated lipid fraction which melted at a higher temperature compared to those of more unsaturated lipids. Lard exhibited lower T_{on} ($-10.30\ ^\circ$C) values, which indicated a higher amount of unsaturated fatty acid.

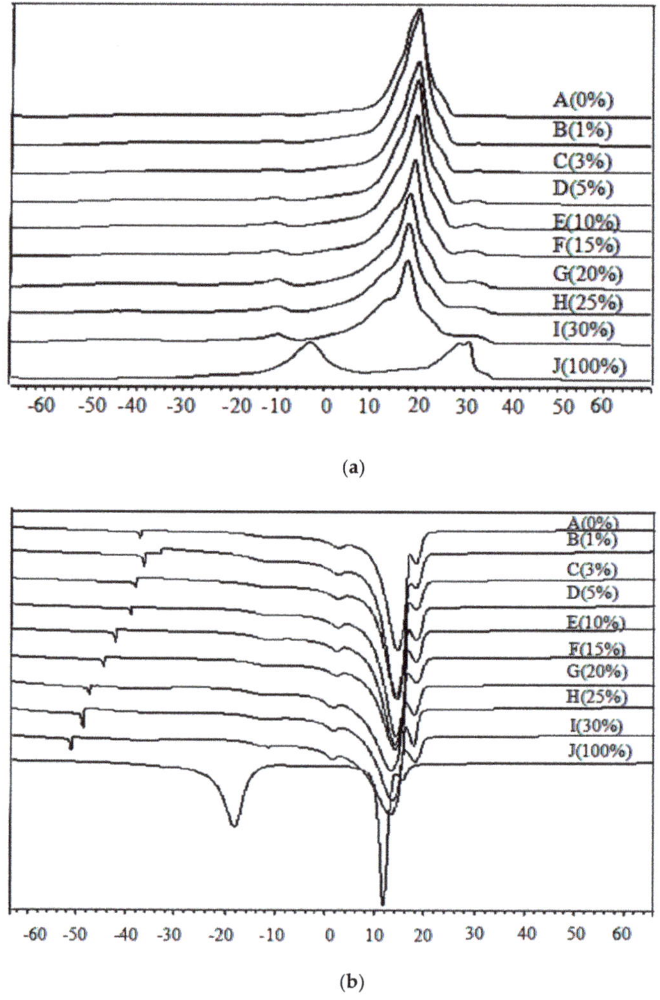

(a)

(b)

Figure 2. Differential scanning calorimetry (**a**) heating and (**b**) cooling thermogram of lard (LD; J: 100%), cocoa butter (CB; A: 0%), and their admixtures.

When cocoa butter is adulterated with lard, new groups of TAG with higher melting points could be introduced into the system, which may eventually change the original profile of cocoa butter. The heating curve of adulterant lard was also found to be completely different from the heating profile of cocoa butter. It was reported that any change in TAG composition influenced the thermal profiles of oils and fats [28]. Animal fats have different levels of saturated and unsaturated fatty acids, even though they have similar physical properties. Yanty et al. [48] reported that the more saturated the TAG, the higher the melting temperature.

There is one major endothermic peak at 20.16 $^\circ$C and two minor endothermic peaks at $-26.00\ ^\circ$C and 34.50 $^\circ$C in cocoa butter adulterated with 3% to 30% lard influenced by the adulterant. As the concentration of lard increased from 3% to 30%, two minor peaks at $-26.00\ ^\circ$C and 34.50 $^\circ$C started to appear and a minor peak at 34.50 °C gradually overlapped with the neighbouring major peak. The major peak at 20.16 °C decreased in size resulting in a broader peak. Peak enthalpy tended to decrease from 84.53 J/g to 72.63 J/g as the adulterant increased from 1% to 30%. The heating profiles gave an indication of the amount of crystallized fat and the occurrence of polymorphic transitions [49]. The DSC heating profile has been used for determining the melting points and various polymorphic forms related to fat crystals. Compositional changes such as fatty acid chain length, the degree of unsaturation, and nature of the distribution of fatty acid in triacylglycerol species has an influence on phase transitions in fats and oils [22]. As reported by Che Man et al. [50], trisaturated TAG (SSS) melted at a higher temperature compared to triunsaturated TAG (UUU), monounsaturated (SSU), and diunsaturated (SUU) triglycerides.

The DSC cooling thermograms of lard, cocoa butter, and their mixture are shown in Figure 2b. The cooling profiles of lard displayed two marked exothermic peaks as reported by Chiavaro et al. [51]. Lard exhibited two major exothermic peaks observed at 17.99 °C and 11.98 °C, which are slightly higher compared to those reported by Nurrulhidayah et al. [26]. This difference could be due to the nature of lard, type of feeding, and variety of fatty acid composition. Cocoa butter exhibited one major exothermic peak at 14.58 °C and three small shoulder peaks appeared at −36.9 °C, 3.16 °C, and 17.97 °C. As reported by Dahimi et al. [52], this cooling behavior of pure samples could be attributed to the amount of saturated and unsaturated TAGs in the samples.

A summary of the cooling properties of lard, cocoa butter, and their mixture can be seen in Table 4. The onset of the cooling (T_{on}) of lard is at −15.00 °C, while that of cocoa butter is at 16.94 °C. The cooling enthalpy for lard (−32.51 J/g) was significantly different ($p < 0.05$) from that of cocoa butter (−85.60 J/g). Lard has a lower cooling enthalpy (−32.51 J/g) due to the presence of free fatty acids and lipid oxidation products; these molecules will be absorbed into the crystal lattices of TAG, forming mixed crystals [53] which require lower enthalpy to undergo phase transition. The unsaturated FA and TAG will crystallize at a low temperature, while the saturated FA and TAG will crystallize at a higher temperature. The melting behavior of edible fats and oils varies due to the different characteristics of FA composition, as reported by Fasina et al. [54].

Table 4. Cooling and heating thermograms of cocoa butter, lard, and their admixtures.

Adulterated Samples (%)	Cooling Properties				Heating Properties			
	Onset (°C)	Offset (°C)	Enthalphy (J/g)	Peak (°C)	Onset (°C)	Offset (°C)	Enthalphy (J/g)	Peak (°C)
0 (CB)	16.94	9.49	−85.60	14.58	14.19	23.38	86.81	20.16
1	16.83	9.37	−85.41	14.41	13.78	23.36	84.53	20.11
3	16.71	9.18	−84.53	14.38	13.60	23.34	78.87	20.09
5	16.64	9.09	−83.55	14.27	13.53	23.21	77.04	20.06
10	16.55	8.47	−82.98	14.08	13.50	23.01	76.41	19.83
15	16.42	7.77	−82.31	14.05	13.48	22.33	75.01	19.43
20	16.26	6.65	−80.64	13.62	13.45	22.26	74.65	18.85
25	15.97	5.73	−79.70	13.30	13.40	21.95	74.15	18.10
30	15.70	5.08	−78.67	13.12	13.37	21.64	72.63	17.78
100 (LD)	−15.00	−22.16	−32.51	11.98	−10.30	2.38	29.86	−4.00

Abbreviations: CB, cocoa butter; LD, lard.

As lard adulteration increases from 1% to 30%, this shoulder peak at 17.97 °C gradually increases in size and shifts to a higher temperature. This shoulder peak is of particular interest due to its sensitivity to lard adulteration. In addition, cooling thermograms of adulterated cocoa butter with 1% to 30% lard showed a minor peak shift to a lower temperature from −36 °C to −41.5 °C. The major peak enthalpy decreased gradually from −85.41 J/g to a lower temperature (−78.67 J/g). Enlargement of the shoulder peaks of cocoa butter and lard which appeared at 17.97 °C could be due to the lower group melting of TAG. It was reported that the adulteration of cocoa butter with lard causes a shift in the peak temperature due to the binary mixture behaviour of the oil samples [21].

4. Conclusions

The results from this study revealed that increasing the lard concentration from 1% to 30% will increase the level of oleic acid (C18:1), while the amount of palmitic acid (C16:0) and stearic acid (C18:0) will decrease. With the addition of lard, the amount of POS, SOS, and POP decreased. An increased lard concentration from 1% to 30%, increased the total triunsaturated TAGs from 0.63% to 3.65%. A minor peak appeared around 34 °C with lard adulteration of 3% to 30%. Cooling thermograms of adulterated cocoa butter with 1% to 30% lard showed a minor peak shifted to a lower temperature of −36 °C to −41.5 °C. Increased lard concentration from 1% to 30% in cocoa butter increased the (C10:0), (C12:0), (C16:1), (C17:0), and (C18:3) fatty acids. The addition of lard increased the amount of LLLn, LLL, and OOL triglycerides. The thermal properties during heating were influenced by triglycerides and fatty acid compositions. Hence, the results from this study could be used as a basis for the identification of lard from other fats in the food authentication process.

Acknowledgments: The authors gratefully acknowledge the financial support received from the Ministry of Higher Education of Malaysia and Universiti Putra Malaysia under the Fundamental Researh Grant Scheme (FRGS, No. 5540047).

Author Contributions: M.A. carried out all the experimental work and also drafted the manuscript. S.A. conceived the study and participated in the experimental design. All authors contributed to the design and interpretation of experimental results, as well as the editing and revision of the manuscript. All authors have read and approved the final manuscript.

Conflicts of Interest: The authors declare no conflict of interest.

References

1. Food and Agriculture Organization (FAO). *Assuring Food Safety and Quality*; Agriculture and Consumer Protection: Rome, Italy, 2014.
2. Che Man, Y.B.; Syaharizaa, Z.A.; Mirghania, M.E.S.; Jinapb, S.; Bakara, J. Analysis of potential lard adulteration in chocolate and chocolate products using Fourier transform infrared spectroscopy. *Food Chem.* **2005**, *90*, 815–819. [CrossRef]
3. Kamal, M.; Karoui, R. Analytical methods coupled with chemometric tools for determining the authenticity and detecting the adulteration of dairy products: A review. *Trends Food Sci. Technol.* **2015**, *46*, 27–48. [CrossRef]
4. Aida, A.A.; Che Man, Y.B.; Wong, C.M.V.L.; Raha, A.R.; Son, R. Analysis of raw meats and fats of pigs using polymerase chain reaction for Halal authentication. *Meat Sci.* **2005**, *69*, 47–52. [CrossRef] [PubMed]
5. Aida, A.A.; Che Man, Y.B.; Raha, A.R.; Son, R. Detection of pig derivatives in food products for Halal authentication by polymerase chain reaction—Restriction fragment length polymorphism. *J. Sci. Food Agric.* **2007**, *87*, 569–572. [CrossRef]
6. Rosman, N.; Mokhtar, N.F.K.; Ali, M.E.; Mustafa, S. Inhibitory effect of chocolate components toward lard detection in chocolate using real time PCR. *Int. J. Food Prop.* **2016**, *19*, 2587–2595. [CrossRef]
7. Che Man, Y.; Mirghani, M. Detection of lard mixed with body fats of chicken,lamb, and cow by fourier transform infrared spectroscopy. *J.Am. Oil Chem. Soc.* **2001**, *78*, 753–761. [CrossRef]
8. Jaswir, I.; Mirghani, M.E.S.; Hassan, T.H.; Mohd Said, M.Z. Determination of lard in mixtures of body fats of mutton and cow by Fourier transform-infra red (FTIR) spectroscopy. *J. Oleo Sci.* **2003**, *52*, 633–638. [CrossRef]
9. Syahariza, Z.A.; Che Man, Y.B.; Selamat, J.; Bakar, J. Detection of lard adulteration in cake formulation by Fourier transform infrared (FTIR) spectroscopy. *Food Chem.* **2005**, *92*, 365–370. [CrossRef]
10. Syahariza, Z.A. Detection of Lard in Selected Food Model Systems Using Fourier Transform Infrared (FTIR) Spectroscopy. Master's Thesis, Universiti Putra Malaysia, Selangor, Malaysia, 2006.
11. Rohman, A.; Che Man, Y.B. FTIR spectroscopy combined with chemometrics for analysis of lard in the mixtures with body fats of lamb, cow, and chicken. *Int. Food Res. J.* **2010**, *17*, 519–526.
12. Rohman, A.; Che Man, Y.B. Analysis of cod-liver oil adulteration using fourier transform infrared (FTIR) spectroscopy. *J. Am. Oil Chem. Soc.* **2009**, *86*, 1149–1153. [CrossRef]

13. Mansor, T.S.T.; Che Man, Y.B.; Rohman, A. Application of fast chromatography and Fourier transform infrared spectroscopy for analysis of lard adulteration in virgin coconut oil. *Food Anal. Methods* **2011**, *4*, 365–372. [CrossRef]

14. Rohman, A.; Che Man, Y.B. Authentication analysis of cod liver oil from beef fat using fatty acid composition and FTIR spectra. *Food Addit. Contam. Part A* **2011**, *28*, 1469–1474. [CrossRef] [PubMed]

15. Che Man, Y.B.; Rohman, A. Differentiation of lard from other edible fats and oils by means of fourier transform infrared spectroscopy and chemometrics. *J. Am. Oil Chem. Soc.* **2011**, *88*, 187–192. [CrossRef]

16. Rohman, A.; Gupitasari, I.; Purwanto, P.; Triyana, K.; Rosman, A.S.; Ahmad, S.A.S.; Yusof, F.M. Quantification of lard in the mixture with olive oil in cream cosmetics based on FTIR spectra and chemometrics. *J. Teknol.* **2014**, *69*, 113–119. [CrossRef]

17. Che Man, Y.B.; Marina, A.M.; Rohman, A.; Al-Kahtani, H.A.; Norazura, O. A FTIR spectroscopy method for analysis of palm oil adulterated with lard in prefried French fries. *Int. J. Food Prop.* **2014**, *17*, 354–362. [CrossRef]

18. Suparman, S.; Rahayu, W.S.; Sundhani, E.; Saputri, S.D. The use of Fourier transform infrared spectroscopy (FTIR) and gas chromatography mass spectroscopy (GCMS) for Halal authentication in imported chocolate with various variants. *J. Food Pharm. Sci.* **2015**, *2*, 6–11.

19. Ramli, S.; Talib, R.A.; Rahman, R.A.; Zainuddin, N.; Othman, S.H.; Rashid, N.M. Detection of lard in ink extracted from printed food packaging using fourier transform infrared spectroscopy and multivariate analysis. *J. Spectrosc.* **2015**, *2015*, 1–6. [CrossRef]

20. Che Man, Y.B.; Gan, H.L.; Nor Aini, I.; Nazimah, S.A.H.; Tan, C.P. Detection of lard adulteration in RBD palm olein using an electronic nose. *Food Chem.* **2005**, *90*, 829–835. [CrossRef]

21. Marikkar, J.M.N.; Lai, O.M.; Ghazali, H.M.; CheMan, Y.B. Detection of lard and randomized lard as adulterants in refined-bleached-deodorized palm oil by differential scanning calorimetry. *J. Am. Oil Chem. Soc.* **2001**, *78*, 1113–1119. [CrossRef]

22. Marikkar, J.M.N.; Ghazali, H.M.; Che Man, Y.B.; Lai, O.M. The use of cooling and heating thermograms for monitoring of tallow, lard and chicken fat adulterations in canola oil. *Food Res. Int.* **2002**, *35*, 1007–1014. [CrossRef]

23. Mansor, T.S.T.; Che Man, Y.B.; Shuhaimi, M. Employment of differential scanning calorimetry in detecting lard adulteration in virgin coconut oil. *J. Am. Oil Chem.Soc.* **2012**, *89*, 485–496. [CrossRef]

24. Marikkar, J.M.N.; Dzulkifly, M.H.; Nadiha, M.Z.N.; Man, Y.B.C. Detection of animal fat contaminations in sunflower oil by differential scanning calorimetry. *Int. J. Food Prop.* **2012**, *15*, 683–690. [CrossRef]

25. Marikkar, J.M.N.; Rana, S. Use of differential scanning calorimetry to detect canola oil (*Brassica napus* L.) adulterated with lard stearin. *J. Oleo Sci.* **2014**, *63*, 867–873. [CrossRef] [PubMed]

26. Nurrulhidayah, A.F.; Arieff, S.R.; Rohman, A.; Amin, I.; Shuhaimi, M.; Khatib, A. Detection of butter adulteration with lard using differential scanning calorimetry. *Int. Food Res. J.* **2015**, *22*, 832–839.

27. Rashood, K.A.; Shaaban, R.R.A.; Moety, E.M.A.; Rauf, A. Compositional and thermal characterization of genuine and randomized lard: A comparative study. *J. Am. Oil Chem. Soc.* **1996**, *73*, 303–309. [CrossRef]

28. Marikkar, J.M.N.; Ghazali, H.M.; Che Man, Y.B.; Peiris, T.S.G.; Lai, O.M. Distinguishing lard from other animal fats in admixtures of some vegetable oils using liquid chromatographic data coupled with multivariate data analysis. *Food Chem.* **2005**, *91*, 5–14. [CrossRef]

29. Marikkar, J.M.N.; Ghazali, H.M.; Long, K.; Lai, O.M. Lard uptake and its detection in selected food products deep-fried in lard. *Food Res. Int.* **2003**, *36*, 1047–1060. [CrossRef]

30. Saeed, T.; Ali, S.C.; Rahman, H.A.; Saway, W.N. Detection of pork and lard as adulterants in processed meat: Liquid chromatographic analysis of derivatized triglycerides. *J. Assoc. Off. Anal. Chem.* **1989**, *72*, 921–925. [PubMed]

31. Farag, R.S.; Abo-raya, S.H.; Ahmed, F.A.; Hewedi, F.M.; Khalifa, H.H. Fractional crystallization and gas chromatographic analysis of fatty acids as a means of detecting butterfat adulteration. *J. Am. Oil Chem. Soc.* **1983**, *60*, 1665–1669. [CrossRef]

32. Farag, R.S.; Ahmed, F.A.; Shihata, J.A.A.; Aboraya, S.H.; Abdalla, A.F. Use of unsaponifiable matter for detection of ghee adulteration with other fats. *J. Am. Oil Chem. Soc.* **1982**, *59*, 557–560. [CrossRef]

33. Dugo, P.; Kumm, T.; Fazio, A.; Dugo, G.; Mondello, L. Determination of beef tallow in lard through a multidimensional off-line non-aqueous reversed phase-argentation LC method coupled to mass spectrometry. *J. Sep. Sci.* **2006**, *29*, 567–575. [CrossRef] [PubMed]

34. Goudjil, H.; Fontecha, J.; Fraga, M.J.; Juarez, M. TAG composition of ewe's milk fat, Detection of foreign fats. *J. Am.Oil Chem. Soc.* **2003**, *80*, 219–222. [CrossRef]

35. Indrasti, D.; Che Man, Y.B.; Mustafa, S.; Hashim, D.M. Lard detection based on fatty acids profile using comprehensive gas chromatography hyphenated with time-of-flight mass spectrometry. *Food Chem.* **2010**, *122*, 1273–1277. [CrossRef]

36. Chin, S.T.; Che Man, Y.B.; Tan, C.P.; Hashim, D.M. Rapid profiling of animal-derived fatty acids using fast GC 3 GC coupled to time-of-flight mass spectrometry. *J. Am. Oil Chem. Soc.* **2009**, *86*, 949–958. [CrossRef]

37. Hoffmann, G. *The Chemistry and Technology of Edible Oils and Fats and Their High Fat Products*; Academic press INC: San Diego, CA, USA, 1989; pp. 1–384.

38. Sionek, B.; Krygier, K.; Ukalski, K.; Ukalska, J.; Amarowicz, R. The influence of nitrogen and carbon dioxide on the oxidative stability of fully refined rapeseed oil. *Eur. J. Lipid Sci. Technol.* **2013**, *115*, 1426–1433. [CrossRef]

39. Haryati, T.; Che Man, Y.B.; Ghazali, H.M.; Asbi, B.A.; Buana, L. Determination of iodine value of palm oil based on triglyceride composition. *J. Am. Oil Chem. Soc.* **1998**, *75*, 789–792. [CrossRef]

40. Marina, A.M.; Che Man, Y.B.; Nazimah, S.A.H.; Amin, I. Monitoring adulteration of virgin coconut oil by selected vegetable oils using differential scanning calorimetry. *J. Food Lipids* **2009**, *16*, 50–61. [CrossRef]

41. Tan, C.P.; Che Man, Y.B. Comparative differential scanning calorimetric analysis of vegetable oils: I. Effects of heating rate variation. *Phytochem. Anal.* **2002**, *13*, 129–141. [CrossRef] [PubMed]

42. Shukla, V.K.L. Cocoa butter properties and quality. *Lipid Technol.* **1995**, *7*, 54–57.

43. Kallio, H.; Yli-Jokipli, K.; Kurvinen, J.P.; Sjovall, O.; Tahvonen, R. Regioisomerism of triacylglycerols in lard, tallow, yolk, chicken skin, palm oil, palm olein, palm stearin, and transesterified blend of palm stearin and coconut oil analyzed by tandem mass spectrometry. *J. Agric. Food Chem.* **2001**, *49*, 3363–3369. [CrossRef] [PubMed]

44. Nizar, N.N.A.; Marikkar, J.M.N.; Hashim, D.M. Differentiation of lard, chicken fat, beef fat and mutton fat by GCMS and EA-IRMS techniques. *J. Oleo Sci.* **2013**, *62*, 459–464. [CrossRef]

45. Cheong, L.Z.; Hong, Z.; Lise, N.; Jensen, K.; Haagensen, J.A.; Xuebing, X. Physical and sensory characteristics of pork sausages from enzymatically modified blends of lard and rapeseed oil during storage. *Meat Sci.* **2010**, *85*, 691–699. [CrossRef] [PubMed]

46. Nurjuliana, M.; Che Man, Y.B.; Hashim, D.M. Analysis of lard's aroma by an electronic nose for rapid halal authentication. *J. Am. Oil Chem. Soc.* **2011**, *88*, 75–82. [CrossRef]

47. Edwards, H.M.J.; Denman, F. Carcass composition studies. 2. Influence of breed, sex and diet on gross composition of the carcass and fatty acid composition of adipose tissue. *Poult. Sci.* **1975**, *54*, 1230–1238. [CrossRef] [PubMed]

48. Yanty, N.A.M.; Marikkar, J.M.N.; Che Man, Y.B.; Long, K. Composition and thermal analysis of lard stearin and lard olein. *J. Oleo Sci.* **2011**, *60*, 333–338. [CrossRef] [PubMed]

49. Fredrick, E.; Foubert, I.; Van De Sype, J.; Dewettinck, K. Influence of monoglycerides on the crystallization behavior of palm oil. *Cryst. Growth Des.* **2008**, *8*, 1833–1839. [CrossRef]

50. Che Man, Y.B.; Hariyati, T.; Ghazali, H.M.; Asbi, B.A. Compositional and thermal profile of crude palm oil and its products. *J. Am. Oil Chem. Soc.* **1999**, *76*, 237–242. [CrossRef]

51. Chiavaro, E.; Vettadini, E.; Estrada, M.T.R.; Cerretani, L.; Bendini, A. Differential scanning calorimeter application to the detectionof refined hazelnut oil in extra virgin olive oil. *Food Chem.* **2008**, *110*, 248–256. [CrossRef] [PubMed]

52. Dahimi, O.; Rahim, A.A.; Abdulkarim, S.M.; Hassan, M.S.; Hashari, S.B.; Mashitoh, A.S. Multivariate statistical analysis treatment of DSC thermal properties for animal fat adulteration. *Food Chem.* **2014**, *158*, 132–138. [CrossRef] [PubMed]

53. Jacobsberg, B.; Ho, O.C. Studies in palm crystallization. *J. Am. Oil Chem. Soc.* **1976**, *53*, 609–617. [CrossRef]

54. Fasina, O.O.; Craig-Schmidt, M.; Colley, Z.; Hallman, H. Predicting melting characteristics of vegetable oils from fatty acid composition. *LWT Food Sci. Technol.* **2008**, *41*, 1501–1505. [CrossRef]

 *foods*

Article

Nutrient and Total Polyphenol Contents of Dark Green Leafy Vegetables, and Estimation of Their Iron Bioaccessibility Using the In Vitro Digestion/Caco-2 Cell Model

Francis Kweku Amagloh [1,*,†], Richard Atinpoore Atuna [2,†], Richard McBride [3,†], Edward Ewing Carey [4,†] and Tatiana Christides [3,†]

1   Food Science & Technology Department, Faculty of Agriculture, University for Development Studies, Nyankpala, Ghana
2   Department of Agricultural Biotechnology and Molecular Biology, Faculty of Agriculture, University for Development Studies, Nyankpala, Ghana; richtuna024@gmail.com
3   Department of Life & Sports Science, Faculty of Engineering and Science, University of Greenwich at Medway, Central Avenue Chatham Maritime, Kent ME4 4TB, UK; richard_mcbride@hotmail.co.uk (R.M.); T.Christides@greenwich.ac.uk (T.C.)
4   International Potato Centre, Kumasi, Ghana; e.carey@cgiar.org
*   Correspondence: fkamagloh@uds.edu.gh; Tel.: +233-0-50-711-3355
†   These authors contributed equally to this work.

Received: 8 May 2017; Accepted: 4 July 2017; Published: 22 July 2017

Abstract: Dark green leafy vegetables (DGLVs) are considered as important sources of iron and vitamin A. However, iron concentration may not indicate bioaccessibility. The objectives of this study were to compare the nutrient content and iron bioaccessibility of five sweet potato cultivars, including three orange-fleshed types, with other commonly consumed DGLVs in Ghana: cocoyam, corchorus, baobab, kenaf and moringa, using the in vitro digestion/Caco-2 cell model. Moringa had the highest numbers of iron absorption enhancers on an "as-would-be-eaten" basis, β-carotene (14169 µg/100 g; $p < 0.05$) and ascorbic acid (46.30 mg/100 g; $p < 0.001$), and the best iron bioaccessibility (10.28 ng ferritin/mg protein). Baobab and an orange-fleshed sweet potato with purplish young leaves had a lower iron bioaccessibility (6.51 and 6.76 ng ferritin/mg protein, respectively) compared with that of moringa, although these three greens contained similar ($p > 0.05$) iron (averaging 4.18 mg/100 g) and β-carotene levels. The ascorbic acid concentration of 25.50 mg/100 g in the cooked baobab did not enhance the iron bioaccessibility. Baobab and the orange-fleshed sweet potato with purplish young leaves contained the highest levels of total polyphenols (1646.75 and 506.95 mg Gallic Acid Equivalents/100 g, respectively; $p < 0.001$). This suggests that iron bioaccessibility in greens cannot be inferred based on the mineral concentration. Based on the similarity of the iron bioaccessibility of the sweet potato leaves and cocoyam leaf (a widely-promoted "nutritious" DGLV in Ghana), the former greens have an added advantage of increasing the dietary intake of provitamin A.

Keywords: β-carotene; Caco-2 cell; iron bioaccessibility; leafy vegetable; polyphenols

---

1. Introduction

It is generally accepted that dark green leafy vegetables (DGLVs) are important sources of micronutrients such as iron and vitamin A. For example, on the basis of compositional data, DGLVs were reported to contribute about 19–39% of iron and 42–68% of vitamin A [1] in the diets of rural South Africans. However, iron and vitamin A deficiencies are perennial malnutrition problems in developing countries where DGLVs are important food ingredients [2,3]. One of the common food ingredients, possibly with a high concentration of micronutrients such as iron and β-carotene (provitamin A), are the greens. However, Cercamondi and co-workers [4] reported that sauce prepared from amaranth (*Amaranthus cruentus*) or Jew's mallow/corchorus (*Corchorus olitorius*) and examples of DGLVs, eaten with a thick maize paste by young Burkinabe women, did not increase the amount of iron absorbed. An inadequate dietary intake of bioavailable iron and vitamin A could be the primary cause of iron and vitamin A deficiencies. Therefore, the bioaccessibility of minerals from food may not solely depend on their concentration, but also on other constituents in the food.

Polyphenols and phytates in cereal and leguminous foods have been shown to limit the bioaccessibility, and consequently, the bioavailability of essential micronutrients including iron, calcium and zinc [5,6]; these staples are usually consumed with DGLVs that may also contain significant levels of these inhibitors. In a human feeding trial conducted by Garcia-Casal and co-workers [7], it was found that β-carotene enhances iron absorption when added to cereal-based diets. This finding was confirmed using Caco-2 cells as a model for iron availability [8]. Thus, the consumption of these greens, reported to be rich in micronutrients such as β-carotene [9,10], should have a double impact as a provitamin A dietary source, and also as an enhancer of iron absorption. However, this was contrary to the findings of Cercamondi and co-workers [4]. This calls for the need to investigate the iron bioaccessibility of commonly consumed DGLVs in Ghana, as anaemia (not categorised) prevalence has consistently been stated to be above 73% for children under 5 years, and at 35% among women of reproductive age in northern Ghana [11–13], where the consumption of greens is high. Vitamin A deficiency among Ghanaian children under 5 years was approximately 79% [14], expectedly, as micronutrient deficiencies usually occur together. In Ghana, DGLVs have been reported to be reliable sources of β-carotene for the majority of the population [10].

Amaranth and jute are widely consumed DGLV in northern Ghana, in addition to others such as baobab (*Adansonia digitata*), and moringa (*Moringa oleifera*) [15]. Sweet potato (*Ipomoea batatas*) is available in northern Ghana [16], but is mainly cultivated for the roots. Sweet potato leaf has been reported to contain appreciable levels of vitamin A, iron and other essential nutrients, including water-soluble vitamins [17,18], and the crop can be cultivated with low agricultural inputs [19]. Also, it has been reported that the sweet potato leaves have higher caffeoylquinic acid derivatives (polyphenols) than commercial vegetables with physiological functions, due to their enhanced antimutagenic and antioxidative properties [20]. Although the polyphenols have health benefits, they may compromise the iron bioaccessibility from the DGLVs. Different polyphenols exist, and have differing effects on the iron bioaccessibility [21–24]. Based on the nutrient superiority of the sweet potato leaf [17], it could serve as an alternative source of leafy vegetables to the populace in tropical regions of the world, particularly in Africa, where vitamin A and iron deficiencies often co-exist and remain public health problems [2,3]. The compositional data suggest that sweet potato and moringa leaves might be better sources of bioavailable iron, compared with other leafy green vegetables, as both have high levels of iron and β-carotene—a dietary factor that has been reported to improve iron bioaccessibility. However, the use of the greens of sweet potato as a leafy vegetable in Ghana is limited.

There is a need to do a comparative study of leaves commonly consumed, and sweet potato leaf before the latter could be suggested as an alternative green in Ghana, as a source of bioavailable iron or β-carotene. The in vitro digestion/Caco-2 cell model has been suggested to be less expensive than human trials [25,26], a more physiological tool for screening iron availability in comparison with solubility and dialysability methods, and an effective approach for predicting the iron bioaccessibility from food for humans [27,28]. Therefore, the in vitro digestion/Caco-2 cell model, with ferritin formation as a marker for iron absorption, was used to measure the iron bioaccessibility of selected greens available in Ghana, in comparison with sweet potato leaves.

The objectives of this study were to compare the nutrient contents and iron bioaccessibility using the in vitro digestion/Caco-2 cell model of five different cultivars of sweet potato, with five other commonly consumed DGLVs in Ghana: cocoyam (*Xanthosoma sagittifolium*), corchorus, baobab, kenaf (*Hibiscus cannabinus*) and moringa.

2. Materials and Methods

2.1. Sample Cultivation and Collection

Five cultivars of sweet potato—three orange-fleshed (Coded OFSP1, OFSP2 and OFSP3), one purple-fleshed (PFSP), and one white-fleshed (WFSP)—and three other DGLVs, namely moringa, corchorus and kenaf (Figures 1 and 2), were nursed in a screen house up to maturity (8 weeks). Each DGLV was cultivated in three replicates, and each replicate contained five pots of the particular green. Baobab and cocoyam were purposively sampled from three different geographical locations. Baobab leaves were collected from trees near settlements from the Upper East, Upper West and North regions, while cocoyam leaves were harvested from farmlands from the Ashanti, East and Brong-Ahafo regions of Ghana. The baobab was not nursed due to a relatively long time for the initiation of vegetative growth. Cocoyam is normally cultivated in the rainforest regions in Ghana and not in northern Ghana.

Figure 1. Cultivars of sweet potato (*Ipomoea batatas*) leaves used in the study. OFSP: orange-fleshed sweet potato; PFSP: purple-fleshed sweet potato; WFSP: white-fleshed sweet potato.

Figure 2. Commonly consumed dark green leafy vegetables (DGLVs) used in the study.

2.2. Sample Preparation

The replicates of the DGLVs were separately washed twice under running tap water and rinsed in distilled water; about two handfuls of DGLVs put into a stainless steel cup with 100 mL of distilled water added were covered with aluminium foil and boiled until soft, for between approximately 15 and 20 min. The cooked DGLVs were allowed to cool, and all the contents of the cup were transferred into coded, transparent, low-density polyethylene zip-lock bags, and stored in a freezer at −18 °C for 2 weeks. Prior to storage in the freezer, about 5 g aliquot portions were taken for moisture determination. The frozen samples were then freeze-dried (TK-118 Vacuum Freeze-Dryer, True Ten Industrial Company Limited, Taichung, Taiwan) for 72 h. The samples were then milled (Thomas Scientific, Dayton Electric Manufacturing Company Limited, Niles, IL, USA) and sieved into fine powder using a 60 mm sieve.

Triplicate aliquots of three-letter-coded powdered samples were couriered to the University of Greenwich at Medway, Chatham-Maritime United Kingdom, and Massey University, Palmerston North, New Zealand, from Ghana. The moisture determination of fresh leaves was performed in Ghana.

2.3. Compositional Analysis

2.3.1. Moisture and Protein

The moisture contents of freshly harvested leaves and cooked leaves were gravimetrically determined using the forced air oven method (AOAC 925.10). For the milled freeze-dried samples, the vacuum oven protocol (AOAC 926.12), as published in the official methods of analysis of AOAC International [29], was used for the moisture determination.

The concentration of nitrogen in the freeze-dried greens was performed by the Dumas method (AOAC 968.06), and a nitrogen-to-protein conversion factor of 6.25 was used to quantify the amount of protein in the leaves on a fee-for-service basis by Massey University Nutrition Laboratory, Palmerston North, New Zealand.

2.3.2. Mineral Analysis: Calcium, Iron, and Zinc

Approximately 0.50 g of the freeze-dried DGLV samples was microwave-digested using an accelerated reaction system (CEM MARS 5H with XP-1500 vessels) for 20 min at 400 psi and 1200 W. Subsequently, calcium, iron and zinc were quantified using an Inductively Coupled Plasma-Optical Emission Spectrometer (ICP-OES, Perkin–Elmer Optima 4300 DV, Perkin–Elmer, Coventry, UK) using protocols as previously described [30]. A certified reference material (ERMCD281, Sigma-Aldrich, UK) was included and run in parallel with the DGLV samples. The data obtained for all three minerals in the reference material were within 5% of the expected values.

2.3.3. β-Carotene

Other researchers have described the extraction and quantification methods used in this study [31]. Averagely, 0.50 g of the freeze-dried samples of the leaves was used for the extraction. A certified reference material (BCR—485, Sigma-Aldrich now Merck, provided to Sigma-Aldrich from the European Commission Joint Research Centre, Institute for Certified Reference Materials and Measurements, Geel, Belgium) was included in three out of the five batches of extraction carried out on DGLV samples. A mean recovery of 128% was obtained for the β-carotene level for the reference material. Therefore, the values obtained for DGLVs were adjusted for a systematic error of 28%.

2.3.4. Ascorbic Acid

The method for vitamin C determination as published by Lee and Coates [32] was carried out by the Massey University Nutrition Laboratory, Palmerston North, New Zealand, on a fee-for-service basis.

2.3.5. Polyphenols

The Folin–Ciocalteu method described by Isabelle and co-workers [33] was used to quantify the total polyphenols in the samples, as gallic acid equivalents. The Nutrition Laboratory, Massey University, New Zealand Palmerston North, New Zealand, carried out the analysis on a fee-for-service basis.

2.4. In Vitro Digestion/Caco-2 Cell Model for Iron Availability

The iron availability from the freeze-dried DGLVs as received from Ghana was assessed using the TC7 Caco-2 cell clone (INSERM U505, Paris, France) from cell passages 42–45 in the in vitro digestion/Caco-2 cell model, as previously described [34], with slight modification. Averagely, 0.5 g rather than 1 g of the sample was weighed for the assessment, as 1 g of the starting material led to a matrix that was too viscous for the multiple mixing and pH adjustments required in this method. Cells were grown in six-well tissue culture plates for the experiments and maintained in DMEM supplemented with 10% v/v foetal bovine serum (FBS). On days 12 and 13, cell media were changed to MEM without FBS, as in the method developed by Glahn [35,36], to ensure low iron media, but optimal expression of Caco-2 cell iron transport proteins [37]. On day 14, foods were subjected to in vitro digestion with a sequential addition of digestive enzymes to mimic exposure to the stomach and small intestine (pepsin at pH 2, followed by bile/pancreatin at pH 7). Digested foods (digestates) and controls, including a blank "No food/added iron" digestate, were then applied to Caco-2 cells through an upper chamber suspended over the plate wells, created using a 15 kD dialysis membrane fitted over a Transwell insert and held in place with a silicon ring. The membrane protected the cells from the digestive enzymes, and also mimicked the gut mucous layer by only allowing soluble iron of a selected size to be available for enterocyte absorption. Cells were treated for 2 h, the digestates were removed, and the cells were returned to the incubator. The cells were harvested for ferritin 24 h after the initiation of the digestive process. Ferritin was measured using a commercial enzyme-linked immunosorbent assay (Spectro ferritin, RAMCO Laboratories Inc., Stafford, TX, USA), and corrected for differing numbers of cell per tissue culture well by measurement of cell protein as an indicator of cell numbers; the cell protein was measured using the Pierce protein bicinchoninic acid assay. Ferritin values were expressed as ng ferritin/mg cell protein.

2.5. Statistical Analysis

The compositional data were converted to an "as-would-be-eaten" basis prior to statistical analysis, using the dry matter content obtained for the cooked samples prior to storage in the freezer. The univariate analysis, followed by Tukey's studentised range test with the significance set at $p < 0.05$, was used for the compositional data. For the in vitro digestion/Caco-2 cell model for iron availability, the data generated were normalised prior to using the general linear model procedure for one-factor analysis, and the results were presented as interval plots of the means with 95% confidence intervals. The Minitab 16.2.2 (Minitab Inc., State College, PA, USA) statistical package was employed for the data analysis.

3. Results

3.1. Compositional Profile

The data in Table 1 is expressed on the as-would-be-eaten basis, with the exception of the moisture value of the freshly harvested leaves. The moisture content of the sweet potato cultivars ranged from 83 to 87 g/100 g, and it was similar to other cultivars cultivated in China [38]. The greens of the sweet potato cultivars were generally not significantly different ($p > 0.05$) from each other for all the components analysed, with the exception of the total polyphenols.

Table 1. Moisture, micronutrient and total polyphenol levels per 100 g in DGLVs on an "as-would-be-eaten" basis [#].

DGLV	Moisture (g) [Y]	Calcium (mg)	Iron (mg)	Zinc (mg)	β-Carotene (µg)	Ascorbic Acid (mg)	Total Polyphenols (mg GAE) [†]
OFSP1	84.09 ± 0.34 [c,d]	95.61 ± 8.01 [c,d]	3.41 ± 0.36 [a,b]	0.44 ± 0.01 [b]	10,533 [a,b]	0.74 ± 0.16 [c]	506.93 ± 86.76 [b]
OFSP2	84.76 ± 0.75 [b,c,d]	81.04 ± 3.24 [d]	1.89 ± 0.29 [b]	0.42 ± 0.02 [b]	8280 [a,b,c]	0.50 ± 0.15 [c]	356.69 ± 79.60 [c]
OFSP3	87.24 ± 0.13 [a]	103.25 ± 2.59 [c,d]	2.58 ± 0.21 [a,b]	0.36 ± 0.03 [b]	7053 [b,c]	0.45 ± 0.07 [c]	336.38 ± 63.15 [c,d,e]
PFSP	84.30 ± 0.26 [c,d]	84.75 ± 8.83 [c,d]	2.04 ± 0.36 [b]	0.44 ± 0.04 [b]	4472 [b,c]	0.48 ± 0.03 [c]	231.44 ± 49.77 [c,d,e]
WFSP	83.91 ± 0.26 [d]	87.02 ± 6.80 [c,d]	3.27 ± 0.34 [a,b]	0.40 ± 0.03 [b]	9501 [a,b,c]	0.34 ± 0.10 [c]	234.86 ± 0.16 [c,d,e]
Baobab	85.97 ± 0.53 [b]	535.63 ± 22.93 [a]	4.59 ± 1.28 [a]	0.65 ± 0.03 [b]	7166 [b,c]	25.50 ± 0.01 [b]	1646.75 ± 69.44 [a]
Cocoyam	85.23 ± 0.64 [b,c]	166.39 ± 15.13 [b]	2.64 ± 0.16 [a,b]	1.49 ± 0.47 [a]	3911 [c]	1.14 ± 0.01 [c]	196.05 ± 10.96 [e]
Corchorus	78.99 ± 0.38 [f]	121.41 ± 3.61 [c]	2.48 ± 0.23 [a,b]	0.45 ± 0.02 [b]	9298 [a,b,c]	3.53 ± 0.58 [c]	337.94 ± 16.44 [c,d,e]
Kenaf	80.68 ± 0.18 [e]	90.24 ± 17.76 [c,d]	2.94 ± 0.25 [a,b]	0.35 ± 0.05 [b]	8959 [a,b,c]	21.79 ± 1.54 [b]	202.42 ± 9.29 [d,e]
Moringa	78.81 ± 0.42 [f]	186.22 ± 23.81 [b]	4.55 ± 1.88 [a]	0.77 ± 0.06 [b]	14,169 [a]	46.30 ± 4.78 [a]	347.38 ± 14.59 [c,d]
p-Value	<0.001	<0.001	0.002	<0.001	<0.001	<0.001	<0.001

[#] Values are means ± standard deviation ($n = 3$), except for the β-carotene value (mean only); values with different letters [a–f] are significantly different ($p < 0.0001$); DGLV—dark green leafy vegetable; OFSP—orange-fleshed sweet potato; PFSP—purple-fleshed sweet potato; and WFSP—white-fleshed sweet potato. [Y] Moisture determined on freshly harvested leaves. [†] GAE—gallic acid equivalents.

OFSP1, Apomuden, a variety being promoted in Ghana because of the β-carotene content in the storage root [39], had approximately 1.7 times more total polyphenols than the other sweet potato cultivars. The leaves of the sweet potato cultivars were not distinctively superior in the levels of the micronutrients analysed, compared with the other DGLVs. However, OFSP1 contained appreciably higher levels of β-carotene (10,533 µg/100 g) and total polyphenols than the other greens, apart from the β-carotene level in moringa (1.3 times more), and the total polyphenols in baobab, which was about thrice higher. Although the roots of the OFSP cultivars are promoted as a dietary source of vitamin A, moringa leaves actually had the highest β-carotene concentration among the DGLVs investigated. Although the WFSP root is devoid of β-carotene [40], the amount of provitamin A in the leaf was more than that in the greens of OFSP2 and OFSP3.

In contrast, among the commonly consumed DGLVs, only baobab leaves contained the highest amount of calcium ($p < 0.001$): on average, about four times more. There was no significant difference in the iron concentration ($p > 0.05$), but the data showed that the iron level in baobab and moringa (4.59 ± 1.28 and 4.55 ± 1.88 mg/100 g, respectively) was higher. Previous data indicated that moringa contained 28.29 ± 0.05 mg/100 g of compositional iron [17], the highest compared with the seven sweet potato varieties in Ghana; the data in this study followed a similar trend.

Three of the DGLVs with notable amounts of ascorbic acid were moringa, baobab, and kenaf. The total polyphenols in baobab was the highest (1646.75 ± 69.44 mg GAE; $p < 0.001$) among all the DGLVs, including the sweet potato cultivars considered in this study. Moringa had a moderate content of total polyphenols, about one-fifth of that in Baobab ($p < 0.05$).

The concentration of zinc in the cocoyam leaf was 1.49 mg/100 g, about thrice more than the average of all the other DGLVs ($p < 0.001$). A similar trend of the zinc data between moringa and the sweet potato cultivars in this study was observed in a previous study in Ghana [17].

Figure 3 shows the crude protein content of all the DGLVs, ranging from 3.62–6.54 g/100 g on the as-would-be-eaten basis. Moringa contained the highest protein (6.54 ± 0.36 g/100 g), and was significantly different ($p < 0.05$) from the next DGLV, baobab (5.67 ± 0.05 g/100 g), which was followed by two cultivars of sweet potato: OFSP1 (5.37 ± 0.04 g/100 g) and OFSP3 (4.99 ± 0.17 g/100 g). The two DGLVs with the lowest protein levels were WFSP (3.87 ± 0.05 g/100 g) and Cocoyam (3.62 ± 0.17 g/100 g). A trend between the protein data for moringa and the sweet potato cultivars was similar to a previous study in Ghana [17].

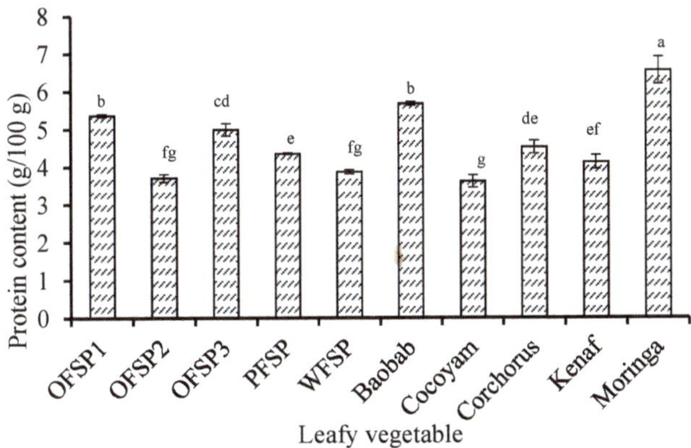

Figure 3. Protein content in "as-would-be-eaten" leafy vegetables. Bar values are means ± standard deviation (n = 3); bars with different letters [a–g] are significantly different ($p < 0.0001$). OFSP—orange-fleshed sweet potato (1, 2 and 3); PFSP—purple-fleshed sweet potato; and WFSP—white-fleshed sweetpotato.

3.2. In Vitro Iron Bioaccessibility Using Caco-2 Cells as a Model

The data representing the in vitro iron bioaccessibility are shown in Figure 4. The overall mean of the iron bioaccessibility was 7.71 ng ferritin/mg protein. Moringa markedly had the best iron bioaccessibility, 10.28 ± 2.73 ng ferritin/mg protein, and was significantly different ($p < 0.0001$) from all the DGLVs investigated.

The two greens (baobab and OFSP1) that could be ranked first and second in terms of the concentrations of total polyphenols had the lowest iron bioaccessibility using the Caco-2 cell model; their bioaccessibility was below the group mean. Conversely, cocoyam had an iron bioaccessibility at the overall mean, although it contained the lowest concentration of polyphenols. Apart from baobab, moringa and OFSP1, all the other DGLVs had a bioaccessibility similar to that of the overall mean.

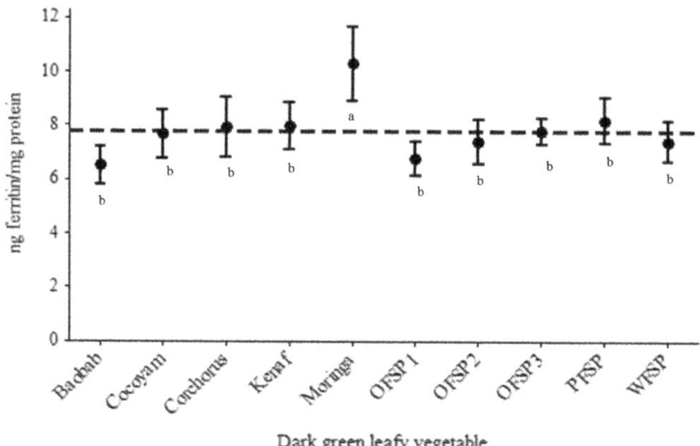

Figure 4. Ferritin formation per half a gram of freeze-dried green leafy vegetables. Vertical lines are means with 95% confidence intervals of ng ferritin/mg protein from the various greens ($n = 12$ for corchorus; $n = 18$ for OFSP1, PFSP, baobab, kenaf and moringa; and $n = 21$ for OFSP2, OFSP3, WFSP and cocoyam) normalised to the blank digest ferritin level; horizontal line indicates the overall mean of ng ferritin/mg protein; means with 95% confidence intervals with a different letter [a,b] are significantly different ($p < 0.0001$).

Table 2. Effect of selected components (on an "as-would-be-eaten" basis) in DGLVs on iron bioaccesibility.

Variable [#]	Estimate (Standard Error)	*p*-Value
Intercept	10.26 (3.24)	0.09
Calcium (mg/100 g)	−0.00 (0.03)	0.99
Iron (mg/100 g)	0.13 (1.18)	0.92
Zinc (mg/100 g)	−0.87 (1.81)	0.68
β-carotene (µg/100 g)	−0.00 (0.00)	0.35
Ascorbic acid (mg/100 g)	0.01 (0.08)	0.93
Total polyphenols (mg GAE/100 g)	−0.00 (0.01)	0.91
Protein (g/100 g)	0.29 (0.78)	0.74

[#] Coefficient of determination ($R^2 = 74.62$).

To test the relationship between the iron bioaccesibility and some of the components (in the DGLVs investigated), a multiple linear regression was conducted (Table 2). Although the model explained about 75% of the variation in the iron bioaccesibility, it was the protein and iron levels that showed a marginal but positive effect such that a unit increase could respectively lead to 0.29 and 0.13 ng ferritin/mg protein formations in Caco-2 cells. However, with regard to zinc, an increase in its concentration resulted in a reduction of the ferritin formation by a 0.87 ng ferritin/mg protein. In this study, ascorbic acid and β-carotene (known enhancers of iron absorption), as well as total polyphenols (inhibitors of iron), had almost no effect on the iron bioaccessibility using the in vitro digestion/Caco-2 cell model.

4. Discussion

OFSP1 was the only sweet potato cultivar with purplish young leaves [41], among the five sweet potato genotypes evaluated in this study. This may have accounted for the highest total polyphenol content of OFSP1, compared to the other sweet potato cultivars. The difference in the iron data for moringa in this study compared to previous work [17] was due to how the data were reported. In the previous study, the result was reported on powdered samples, while in our study, it was on an as-would-be-eaten basis. Nonetheless, the trend of iron levels being the highest in moringa was also confirmed in this study.

Although cocoyam leaf is widely consumed, and promoted in Ghana as a "nutritious" green to improve iron status (anecdotally), on the basis of its composition data, it was highest only in zinc, and lowest in β-carotene and total polyphenols, compared with the OFSP cultivars. Because both the sweet potato leaves and cocoyam had a similar iron bioaccessibility, the sweet potato leaves could be used in culinary preparations, and had an added advantage of increasing the dietary intake of β-carotene, compared to those of cocoyam.

Generally, the level of iron bioaccessibility from the DGLVs was relatively low (6–10 ng ferritin/mg protein) compared with our previous work on complementary food from the same laboratory (12–34 ng ferritin/mg protein) [34]. However, a strong comparison cannot be made between the data from the two studies, as different sample weights were used: 1 g in the previous work, and 0.5 g in the present study. A previous community-based feeding trial using Weanimix, which had an iron bioaccessibility of 17.32 ± 2.84 ng ferritin/mg protein [34], resulted in a poor iron status among older infants in Ghana [42,43]. The lower availability of iron in the greens in this study lends support to the finding of the work on young Burkinabe women, which resulted in no increase in iron absorption after eating Jew's mallow with a thick maize paste [4].

As mentioned earlier, moringa contained the highest number of enhancers of iron absorption: β-carotene [7,8] and ascorbic acid [44]; although in this study their effect were not realised except for the concentration of iron. Additionally, the concentration of total polyphenols in this DGLV was moderate. The composition of nutrients in moringa, compared with the other DGLVs, may have contributed to it having the highest bioaccessibility of iron, as obtained from the in vitro Caco-2 cells model study. Although OFSP1 had significantly similar levels of β-carotene and iron to moringa, and one-third of the total polyphenols of baobab, its iron bioaccessibility was lower than for moringa, indicating that the reported caffeoylquinic acid derivatives in sweet potato leaves [20] may have limited the bioaccessibility of iron. Baobab had the lowest iron bioaccessibility, in spite of being one of the greens that contained the highest amounts of iron and ascorbic acid. This may have been attributed to the high concentration of total polyphenols [20], and not calcium, which is known to inhibit iron absorption [25,45]; and relative to the other DGLVs, suggesting that the polyphenols in baobab may be very inhibitory, even in the presence of endogenous ascorbic acid. However, the amount of calcium in the greens explicitly did not suggest inhibitory effects on iron, as moringa contained the second highest level of this mineral among all the DGLVs investigated, but had a markedly better iron availability. Therefore, predicting iron bioaccessibility based only on compositional data could lead to false conclusions.

The effect of the concentration of plant protein on the iron bioaccessibility cannot be explicitly substantiated in this study. Moringa, having the highest as-would-be-eaten protein, was the green with the highest bioaccessibility. Both baobab and OFSP1, which contained relatively high concentrations of protein compared to the rest of the DGLVs with the exception of Moringa, were those that recorded the lowest bioaccessibility of iron, although not significantly. Thus, from the data in this study, it is difficult to use the protein concentration to predict the iron bioaccessibility, although the regression analysis showed a direct effect. The inverse association between the zinc concentration and the index of iron bioaccessibility could be attributed to the cocoyam leaf, which had the highest zinc concentration and the lowest ferritin formation in the Caco-2 cells.

The major limitations of this study were that phytate and the constituents of the different classes of polyphenols were not quantified. The assay method previously used for phytate determination [46,47] gave very inconsistent results within replicates in this study; possibly the colour of DGLVs interfered with the spectrophotometer readings.

5. Conclusions

The studied greens varied in terms of calcium, iron and zinc levels. In addition, moringa had the highest levels of β-carotene and ascorbic acid. Baobab had the highest levels of calcium and total polyphenols. Within the limits of this study, iron bioaccessibility is influenced by a complex interplay of several components in DGLVs, including protein, ascorbic acid, β-carotene and total polyphenols. Moringa had the best iron bioaccessibility, and the lowest was found in baobab and one of the orange-fleshed sweet potatoes with purplish young leaves. Estimating iron bioaccessibility in greens based on the mineral concentration may lead to incorrect conclusions. Based on the similarity of the iron bioaccessibility of the sweet potato leaves and cocoyam leaf, the widely promoted "nutritious" DGLVs in Ghana, the former greens have an added advantage of increasing the dietary intake of provitamin A.

Acknowledgments: The Nutricia Research Foundation, the Netherlands, awarded FKA an International Training Fellowship (Project number: 2014-T2) to travel to the University of Greenwich, United Kingdom, to conduct some of the laboratory analyses reported in this study. Funding received from the International Potato Center under the project SASHA II: Sweetpotato Action for Security and Health in Africa is heartily valued and acknowledged. The Bill and Melinda Gates Foundation is also acknowledged for paying for Open Access publication.

Author Contributions: Francis Kweku Amagloh, Richard Atinpoore Atuna, Richard McBride, Edward Ewing Carey and Tatiana Christides were involved in the experimental design, data collection and analysis, and approved the final version of the manuscript.

Conflicts of Interest: The authors declare no conflict of interest.

References

1. Faber, M.; Van Jaarsveld, P.; Laubscher, R. The contribution of dark-green leafy vegetables to total micronutrient intake of two-to five-year-old children in a rural setting. *Water SA* **2007**, *33*, 407–412.

2. De Benoist, B.; McLean, E.; Egli, I.; Cogswell, M. *Worldwide Prevalence of Anaemia 1993–2005: WHO Global Database on Anaemia*; WHO Press: Geneva, Switzerland, 2008.

3. World Health Organization. *Global Prevalence of Vitamin A Deficiency in Populations at Risk 1995–2005: WHO Global Database on Vitamin A Deficiency*; WHO Press: Geneva, Switzerland, 2009.

4. Cercamondi, C.I.; Icard-Vernière, C.; Egli, I.M.; Vernay, M.; Hama, F.; Brouwer, I.D.; Zeder, C.; Berger, J.; Hurrell, R.F.; Mouquet-Rivier, C. A higher proportion of iron-rich leafy vegetables in a typical Burkinabe maize meal does not increase the amount of iron absorbed in young women. *J. Nutr.* **2014**, *144*, 1394–1400. [PubMed]

5. Davies, N.T.; Reid, H. An evaluation of the phytate, zinc, copper, iron and manganese contents of, and Zn availability from, soya-based textured-vegetable-protein meat-substitutes or meat-extenders. *Br. J. Nutr.* **1979**, *41*, 579–589. [PubMed]

6. Gautam, S.; Platel, K.; Srinivasan, K. Promoting influence of combinations of amchur, β-carotene-rich vegetables and Allium spices on the bioaccessibility of zinc and iron from food grains. *Int. J. Food Sci. Nutr.* **2011**, *62*, 518–524. [PubMed]

7. Garcia-Casal, M.N.; Layrisse, M.; Solano, L.; Baron, M.A.; Arguello, F.; Llovera, D.; Ramirez, J.; Leets, I.; Tropper, E. Vitamin A and beta-carotene can improve nonheme iron absorption from rice, wheat and corn by humans. *J. Nutr.* **1998**, *128*, 646–650. [PubMed]

8. Garcia-Casal, M.N.; Leets, I. Carotenoids, but not vitamin A, improve iron uptake and ferritin synthesis by Caco-2 cells from ferrous fumarate and NaFe-EDTA. *J. Food Sci.* **2014**, *79*, H706–H712. [PubMed]

9.   Van Jaarsveld, P.; Faber, M.; van Heerden, I.; Wenhold, F.; van Rensburg, W.J.; van Averbeke, W. Nutrient content of eight African leafy vegetables and their potential contribution to dietary reference intakes. *J. Food Compos. Anal.* **2014**, *33*, 77–84.

10.   Takyi, E.E.K. Children's consumption of dark green, leafy vegetables with added fat enhances serum retinol. *J. Nutr.* **1999**, *129*, 1549–1554. [PubMed]

11.   ORC Macro. Ghana Demographic and Health Survey 2003. Available online: http://www.dhsprogram.com/pubs/pdf/FR152/FR152.pdf (accessed on 7 July 2017).

12.   Ghana Statistical Service (GSS); Ghana Health Service (GHS); ICF International. Ghana Demographic and Health Survey 2014. Available online: https://dhsprogram.com/pubs/pdf/FR307/FR307.pdf (accessed on 7 July 2017).

13.   Ghana Statistical Service (GSS); Ghana Health Service (GHS); ICF Macro. Ghana Demographic and Health Survey 2008. Available online: http://www.dhsprogram.com/pubs/pdf/FR221/FR221[13Aug2012].pdf (accessed on 7 July 2017).

14.   World Health Organization. WHO Global Database on Vitamin A Deficiency. Available online: http://www.who.int/vmnis/vitamina/data/database/countries/gha_vita.pdf (accessed on 1 August 2016).

15.   Amagloh, F.K.; Nyarko, E.S. Mineral nutrient content of commonly consumed leafy vegetables in northern Ghana. *Afr. J. Food Agric. Nutr. Dev.* **2012**, *12*, 6397–6408.

16.   Dittoh, S. Improving availability of nutritionally adequate and affordable food supplies at community levels in West Africa. In *International Workshop: Food-Based Approaches for a Healthy Nutrition in West Africa, Proceedings of the 2nd International Workshop, Ouagadougou, Burkina Faso, 23–28 November 2003*; Presses Universitaires de Ouagadougou: Ouagadougou, Burkina Faso, 2004; pp. 23–28.

17.   Oduro, I.; Ellis, W.; Owusu, D. Nutritional potential of two leafy vegetables: *Moringa oleifera* and *Ipomoea batatas* leaves. *Sci. Res. Essays* **2008**, *3*, 57–60.

18.   Barrera, W.A.; Picha, D.H. Ascorbic acid, thiamin, riboflavin, and vitamin B6 contents vary between sweetpotato tissue types. *HortScience* **2014**, *49*, 1470–1475.

19.   Faber, M.; Laurie, S.M.; van Jaarsveld, P.J. Total β-carotene content of orange sweetpotato cultivated under optimal conditions and at a rural village. *Afr. J. Biotechnol.* **2013**, *12*, 3947–3951.

20.   Islam, S. Antimutagenicity and antioxidant activity in the Ipomoea batatas L. genotypes in relation to polyphenolics. In Proceedings of the International Conference on Advances in Agricultural, Biological & Environmental Sciences, London, UK, 22–23 July 2015; pp. 1–7.

21.   Tako, E.; Beebe, S.; Reed, S.; Hart, J.; Glahn, R. Polyphenolic compounds appear to limit the nutritional benefit of biofortified higher iron black bean (*Phaseolus vulgaris* L.). *Nutr. J.* **2014**, *13*, 28. [CrossRef] [PubMed]

22.   Tako, E.; Reed, S.M.; Budiman, J.; Hart, J.J.; Glahn, R.P. Higher iron pearl millet (*Pennisetum glaucum* L.) provides more absorbable iron that is limited by increased polyphenolic content. *Nutr. J.* **2015**, *14*. [CrossRef] [PubMed]

23.   Petry, N.; Egli, I.; Campion, B.; Nielsen, E.; Hurrell, R. Genetic reduction of phytate in common bean (*Phaseolus vulgaris* L.) seeds increases iron absorption in young women. *J. Nutr.* **2013**, *143*, 1219–1224. [CrossRef] [PubMed]

24.   Abizari, A.-R.; Moretti, D.; Schuth, S.; Zimmermann, M.B.; Armar-Klemesu, M.; Brouwer, I.D. Phytic acid-to-iron molar ratio rather than polyphenol concentration determines iron bioavailability in whole-cowpea meal among young women. *J. Nutr.* **2012**, *142*, 1950–1955. [CrossRef] [PubMed]

25.   Glahn, R.P.; Rassier, M.; Goldman, M.I.; Lee, O.A.; Cha, J. A comparison of iron availability from commercial iron preparations using an in vitro digestion/Caco-2 cell culture model. *J. Nutr. Biochem.* **2000**, *11*, 62–68. [CrossRef]

26.   Kamiloglu, S.; Capanoglu, E.; Grootaert, C.; Van Camp, J. Anthocyanin Absorption and Metabolism by Human Intestinal Caco-2 Cells—A Review. *Int. J. Mol. Sci.* **2015**, *16*, 21555–21574. [CrossRef] [PubMed]

27.   Fairweather-Tait, S.; Lynch, S.; Hotz, C.; Hurrell, R.F.; Abrahamse, L.; Beebe, S.; Bering, S.; Bukhave, K.; Glahn, R.; Hambidge, M.; et al. The usefulness of in vitro models to predict the bioavailability of iron and zinc: A consensus statement from the HarvestPlus expert consultation. *Int. J. Vitam. Nutr. Res.* **2005**, *75*, 371–374. [CrossRef] [PubMed]

28.   Tako, E.; Bar, H.; Glahn, R. The combined application of the Caco-2 cell bioassay coupled with in vivo (*Gallus gallus*) feeding trial represents an effective approach to predicting Fe bioavailability in humans. *Nutrients* **2016**, *8*, 732. [CrossRef] [PubMed]

29. Association of Official Analytical Chemists (AOAC). *Official Methods of Analysis of AOAC International*, 18th ed.; AOAC International: Gaithersburg, MD, USA, 2005.

30. Zand, N.; Chowdhry, B.Z.; Wray, D.S.; Pullen, F.S.; Snowden, M.J. Elemental content of commercial 'ready to-feed' poultry and fish based infant foods in the UK. *Food Chem.* **2012**, *135*, 2796–2801. [CrossRef] [PubMed]

31. Bechoff, A.; Westby, A.; Owori, C.; Menya, G.; Dhuique-Mayer, C.; Dufour, D.; Tomlins, K. Effect of drying and storage on the degradation of total carotenoids in orange-fleshed sweetpotato cultivars. *J. Sci. Food Agric.* **2010**, *90*, 622–629. [CrossRef] [PubMed]

32. Lee, H.S.; Coates, G.A. Liquid chromatographic determination of vitamin C in commercial Florida citrus juices. *J. Micronutr. Anal.* **1987**, *3*, 199–209.

33. Isabelle, M.; Lee, B.L.; Lim, M.T.; Koh, W.-P.; Huang, D.; Ong, C.N. Antioxidant activity and profiles of common vegetables in Singapore. *Food Chem.* **2010**, *120*, 993–1003. [CrossRef]

34. Christides, T.; Amagloh, F.K.; Coad, J. Iron bioavailability and provitamin A from sweet potato- and cereal-based complementary foods. *Foods* **2015**, *4*, 463–476. [CrossRef] [PubMed]

35. Glahn, R.P.; Lee, O.A.; Yeung, A.; Goldman, M.I.; Miller, D.D. Caco-2 cell ferritin formation predicts nonradiolabeled food iron availability in an in vitro digestion Caco-2 cell culture model. *J. Nutr.* **1998**, *128*, 1555–1561. [PubMed]

36. Yun, S.M.; Habicht, J.P.; Miller, D.D.; Glahn, R.P. An in vitro digestion/Caco-2 cell culture system accurately predicts the effects of ascorbic acid and polyphenolic compounds on iron bioavailability in humans. *J. Nutr.* **2004**, *134*, 2717–2721. [PubMed]

37. Sharp, P.; Tandy, S.; Yamaji, S.; Tennant, J.; Williams, M.; Singh Srai, S.K. Rapid regulation of divalent metal transporter (DMT1) protein but not mRNA expression by non-haem iron in human intestinal Caco-2 cells. *FEBS Lett.* **2002**, *510*, 71–76. [CrossRef]

38. Sun, H.; Mu, T.; Xi, L.; Zhang, M.; Chen, J. Sweet potato (*Ipomoea batatas* L.) leaves as nutritional and functional foods. *Food Chem.* **2014**, *156*, 380–389. [CrossRef] [PubMed]

39. Islam, S.N.; Nusrat, T.; Begum, P.; Ahsan, M. Carotenoids and β-Carotene in orange fleshed sweet Potato: A possible solution to vitamin A deficiency. *Food Chem.* **2016**, *199*, 628–631. [CrossRef] [PubMed]

40. Van Jaarsveld, P.J.; Faber, M.; Tanumihardjo, S.A.; Nestel, P.; Lombard, C.J.; Benade, A.J.S. β–carotene-rich orange-fleshed sweet potato improves the vitamin A status of primary school children assessed with the modified-relative-dose-response test. *Am. J. Clin. Nutr.* **2005**, *81*, 1080–1087. [PubMed]

41. Tumwegamire, S.; Mwanga, R.O.M.; Andrade, M.; Low, J.W.; Kapinga, R.E.; Ssemakula, G.N.; Laurie, S.M.; Chipungu, F.P.; Ndirigue, J.; Agili, S.; et al. *Orange-Fleshed Sweetpotato for Africa: Catalogue 2014*, 2nd ed.; International Potato Center (CIP): Lima, Peru, 2014.

42. Lartey, A.; Manu, A.; Brown, K.H.; Peerson, J.M.; Dewey, K.G. A randomized, community-based trial of the effects of improved, centrally processed complementary foods on growth and micronutrient status of Ghanaian infants from 6 to 12 mo of age. *Am. J. Clin. Nutr.* **1999**, *70*, 391–404. [PubMed]

43. Lartey, A.; Manu, A.; Brown, K.H.; Peerson, J.M.; Dewey, K.G. Predictors of growth from 1 to 18 months among breast-fed Ghanaian infants. *Eur. J. Clin. Nutr.* **2000**, *54*, 41–49. [CrossRef] [PubMed]

44. Teucher, B.; Olivares, M.; Cori, H. Enhancers of iron absorption: Ascorbic acid and other organic acids. *Int. J. Vitam. Nutr. Res.* **2004**, *74*, 403–419. [CrossRef] [PubMed]

45. Hallberg, L.; Brune, M.; Erlandsson, M.; Sandberg, A.-S.; Rossander-Hulten, L. Calcium: Effect on different amounts on non heme-iron and heme-iron absorption in humans. *Am. J. Clin. Nutr.* **1991**, *53*, 112–119. [PubMed]

46. Amagloh, F.K.; Brough, L.; Weber, J.L.; Mutukumira, A.N.; Hardacre, A.; Coad, J. Sweetpotato-based complementary food would be less inhibitory on mineral absorption than a maize-based infant food assessed by compositional analysis. *Int. J. Food Sci. Nutr.* **2012**, *63*, 957–963. [CrossRef] [PubMed]

47. Amagloh, F.K.; Coad, J. Orange-fleshed sweet potato-based infant food is a better source of dietary vitamin A than a maize-legume blend as complementary food. *Food Nutr. Bull.* **2014**, *35*, 51–59. [CrossRef] [PubMed]

 *foods*

Article

Glucose Content and In Vitro Bioaccessibility in Sweet Potato and Winter Squash Varieties during Storage

Fernanda Zaccari [1,2,*]**, María Cristina Cabrera** [2,3] **and Ali Saadoun** [3]

[1] Poscosecha de Frutas y Hortalizas, Facultad de Agronomía, Universidad de la República, Av. Eugenio Garzón 780, CP12900 Montevideo, Uruguay

[2] Nutrición y Calidad de Alimentos, Facultad de Agronomía, Universidad de la República, Av. Eugenio Garzón 780, CP12900 Montevideo, Uruguay; mcab@fagro.edu.uy

[3] Fisiología y Nutrición, Facultad de Ciencias, Universidad de la República, Iguá 2245, CP11400 Montevideo, Uruguay; asaadoun@fcien.edu.uy

* Correspondence: fzaccari@fagro.edu.uy; Tel.: +598-2359-7191 (ext. 213); Fax: +598-2354-2052

Academic Editor: Theo H. Varzakas

Received: 30 May 2017; Accepted: 27 June 2017; Published: 30 June 2017

Abstract: Glucose content and in vitro bioaccessibility were determined in raw and cooked pulp of Arapey, Cuabé, and Beauregard sweet potato varieties, as well as Maravilla del Mercado and Atlas winter squash, after zero, two, four, and six months of storage (14 °C, 80% relative humidity (RH)). The total glucose content in 100 g of raw pulp was, for Arapey, 17.7 g; Beauregard, 13.2 g; Cuabé, 12.6 g; Atlas, 4.0 g; and in Maravilla del Mercado, 4.1 g. These contents were reduced by cooking process and storage time, 1.1 to 1.5 times, respectively, depending on the sweet potato variety. In winter squash varieties, the total glucose content was not modified by cooking, while the storage increased glucose content 2.8 times in the second month. After in vitro digestion, the glucose content released was 7.0 times higher in sweet potato (6.4 g) than in winter squash (0.91 g) varieties. Glucose released by in vitro digestion for sweet potato stored for six months did not change, but in winter squashes, stored Atlas released glucose content increased 1.6 times. In conclusion, in sweet potato and winter squash, the glucose content and the released glucose during digestive simulation depends on the variety and the storage time. These factors strongly affect the supply of glucose for human nutrition and should be taken into account for adjusting a diet according to consumer needs.

Keywords: glucose; bioaccessibility; *Ipomoea batatas*; *Cucurbita* sp.; postharvest

---

1. Introduction

Sweet potatoes (*Ipomoea batatas*, L.) and winter squashes (*Cucurbita* sp.) are two of the most important starch vegetable crops in the world [1–4]. Both vegetables—originating in the regions of South and Central America and produced and widely consumed in others countries, particularly in parts of Asia, Sub-Saharan Africa, and the Pacific Islands—are necessary for human nutrition [2–4]. The principal components in the fresh roots of sweet potatoes and winter squash fruits are carbohydrates; these make up around 25–28% fresh basis weight for sweet potato pulp, and 5–7% in winter squash [3–6]. Starch is the predominant carbohydrate, making up 85–95% and 11–62 % of total dry matter for sweet potato and winter squash, respectively [3–5]. Both vegetables can be harvested during summer and autumn seasons. The type of starch reserves in these vegetables influence the time of the postharvest conservation, i.e., three months for a medium length of time, or eight months for a long time, depending on the variety and storage conditions [3–5]. The starch, in roots and fruit, is metabolized to simple sugars, such as glucose, and it is used to maintain their viability during the postharvest life [3–7]. The interaction of the varieties, handling during the growing crop, and harvest and storage conditions could change the physiological process in vegetables and, in addition, the process of cooking could directly impact the amount and bioaccessibility of glucose content and of other compositional components of the pulp [7–11]. In spite of the fact that glucose is the key energy for life, particularly for the brain [12–14], a strong relationship between glucose the development of some diseases, such as diabetes, hypertension, and cardiovascular diseases, has been reported [15]. Consequently, sweet potato roots and winter squash fruits largely used in South American crops are interesting as energy sources for children and the elderly, and for persons with chronic diseases such as diabetes. In this last case, an accurate knowledge about glucose content and how much is released during storage or cooking is necessary. For this reason, the aim of this work was to determine the total glucose content and the effect of the cooking process on sweet potato and winter squash varieties during the storage period. Additionally, the glucose release during in vitro digestion was determined in cooked pulp from all varieties and storage times.

2. Experimental Section

2.1. Plant Materials and Sample Preparation

Three sweet potato varieties (Arapey, Cuabé, and Beauregard) are local varieties obtained in a breeding program [16], and two varieties of winter squash, one, a hybrid between *Cucurbita maxima* and *Cucurbita moschata* (named Maravilla del Mercado, Sakata) and the other type, "butternut", *Cucurbita moschata* (Atlas, Sakata), were used. The roots of sweet potatoes and fruits of winter squashes were harvested at the mature stage at the end of summer (April). All of them were maintained for two weeks in an open room under initial handling wound healing conditions. After this time, the roots and fruits were carried to the Postharvest Fruits and Vegetables Laboratory of the Faculty of Agronomy. Selected roots and fruits were stored in a cold room at 14 °C and 80% relative humidity (RH) for zero (harvest), two, four, and six months. They were kept in three randomized plots per variety and storage period with 5–8 roots of sweet potato ($\approx$2 kg per plots) and 12–15 fruits of winter squash per plot ($\approx$25 kg per plots). After each storage time, the fruits were washed with tap water and soft brushing, rinsed with distilled water, drained and dried with blotting paper. Five roots and fruits per plot with no visible defects were used. The equatorial central part of the roots and fruits were used. Butternut squash fruits are pear-shaped. Therefore, in this case, the slices were obtained within the equatorial zone between the stem and the start of the seed cavity of the fruit. The pieces were peeled and cut with a stainless steel knife in cubes with 5 cm and 3 cm side lengths for winter squash pulp or sweet potato, respectively. The cubes were kept in sealed bags in a freezer (-20 °C) until analysis; previously, half of them were cooked in an oven microwave (Kassel®, KS-MM20, Hamburg, Germany) at 800 W, with hot water ($\approx$55 °C) at a ratio of 1:2 (weight pulp: volume water) for 6 min. The final cooked temperature in the pulp was measured (60–65 °C) and the cube had an edible, firm texture and flavor. This cooking process was tested in previous trials. The variables studied were determined in duplicate for each treatment plot following analysis methodologies described below.

2.2. Total Glucose Content in Raw and Cooked Pulp

Extraction of glucose, in raw and cooked pulp, was performed with 0.5 g of pulp in 8 mL of HCl (4 N) boiling for 2 h. The extraction was filtered and NaOH (2 N) was added to neutralize the filtered solution. Total glucose content was measured by colorimetric methods using a commercial enzymatic procedure from Spinreacts kits (Glucose–TR, GOD-PROD, Sant Esteves de Bas, Girona, Spain). Determinations were obtained on a visible spectrophotometer (Genesys 10 VIS; Thermo Electro Corporation, Berlin, Germany) at $\lambda = 505$ nm. Data were expressed in grams of glucose per 100 g fresh weight (g 100 g^{-1} fw).

2.3. In Vitro Digestion of Cooked Pulp

An in vitro model based on a simulation digestion was performed. Samples of cooked pulp from sweet potato and winter squash varieties from every storage time were digested as described by Zaccari et al. [17]. Minor modifications was performed for the duodenal phase, including the addition of 0.1 mL α-amylase (A3306 Sigma-Aldrich, Saint Louis, MO, USA) and 30 µL α-amyloglucosidase (AT7095 Sigma-Aldrich, Saint Louise, MO, USA) at pH 6, and previously adjusting the digests with citrate buffer solution (pH 5.5) and $NaHCO_3$ (0.8 M). After in vitro digestion, the digests were filtered and the glucose content in the extraction was measured.

2.4. In Vitro Bioaccessible Glucose Content in Cooked Pulp

The glucose content in the extracted digests was measured with similar procedures as those described for the total glucose measurement in pulp. Data were expressed in grams of total glucose released by in vitro digestion per 100 g cooked pulp weight (g 100 g^{-1} fw), and the percentage of bioaccessible glucose was calculated as:

$$\% \text{ glucose bioaccessible} = \frac{\text{glucose released in vitro digestion}}{\text{total glucose in cooked pulp}} \times 100 \qquad (1)$$

2.5. Experimental Design and Statistical Analysis

The experimental design had completely randomized plots ($n = 3$), in a $4 \times 2 \times 3$ or 2 with factorial structure, storage factor with four storage times (zero, two, four, and six months), two preparation processes (raw or cooked), with three or two varieties of sweet potato or winter squash, respectively. For sweet potatoes and winter squash, data were analyzed by a three way-ANOVA ($p \leq 0.05$) with variables including varieties, preparation, and storage times. Each preparation process was analyzed by one way-ANOVA ($p \leq 0.05$) for sweet potato or winter squash, for varieties and storage time, followed by a Tukey post-hoc test ($p \leq 0.05$). The effects of the preparation (raw and cooked) on each storage time and variety were analyzed by Student's t test ($p \leq 0.05$). The data was processed with the InfoStat (Version 2015; FCA, Córdoba, Argentine) statistical program. All values were presented as means $\pm$ SEM and expressed per 100 g of fresh pulp weight (100 g^{-1} fw).

3. Results and Discussion

3.1. Total Glucose Content in Raw and Cooked Pulp

The total glucose content in raw and cooked pulp was 3–4-fold higher in sweet potato (Arapey, 17.7 g; Beauregard, 13.2 g; Cuabé, 12.6 g) than in winter squash varieties (Atlas, 4.0 g; Maravilla del Mercado, 4.1 g) (Figure 1). In sweet potato, the total content of glucose depended on the interaction of the variety, method of preparation, and storage time; it decreased with the cooking process and times of storage (Figure 1). In sweet potato, raw and cooked pulp, Arapey was the variety that had the greatest total glucose content during four months of storage, and it decreased around 26% at six months. Furthermore, Beauregard and Cuabé varieties had similar glucose content (14.1 g) until the second month of storage, after which it decreased (10.3 g). The cooking process determined the reduction of the total glucose in the three varieties of sweet potatoes, especially at the second and fourth months of storage in Cuabé and Beauregard varieties (Figure 1).

However, in winter squash pulp, the total glucose content was similar in both varieties (4.0 g) and depended on the preparation and storage time (Figure 1). In raw and cooked pulp, it was observed that the total glucose content increased around 50–55% in the second month of storage. The raw pulp of winter squash had more (4.5 g) glucose content than cooked pulp (3.6 g). The effects of the cooking process were observed only in Maravilla del Mercado at six months of storage, with 50% less glucose than in cooked pulp (Figure 1). Starch and sugars are the main components of the dry matter (40–85%), and the source of glucose reported for both sweet potato and winter squash pulp [3–5,7–9]. Several authors have reported that in varieties of sweet potato and winter squash, losses of dry matter, total sugars, and glucose content for metabolism respiration and degradation during postharvest storage [6,7,9,18–20] were observed. Other authors have determined differences in starch granules as well as the amount and activity of enzyme α and β-amylases, which explain part of the different behaviors between varieties and the decreased rate of the total glucose content during storage time [6,9,11,18,21–24]. For winter squash, an increase in the total glucose content at the second month of storage was observed. This increase in glucose can be explained by the biosynthesis from other compounds and/or by a translocation from other parts of the fruit, as reported for watermelon [25] and melon [26].

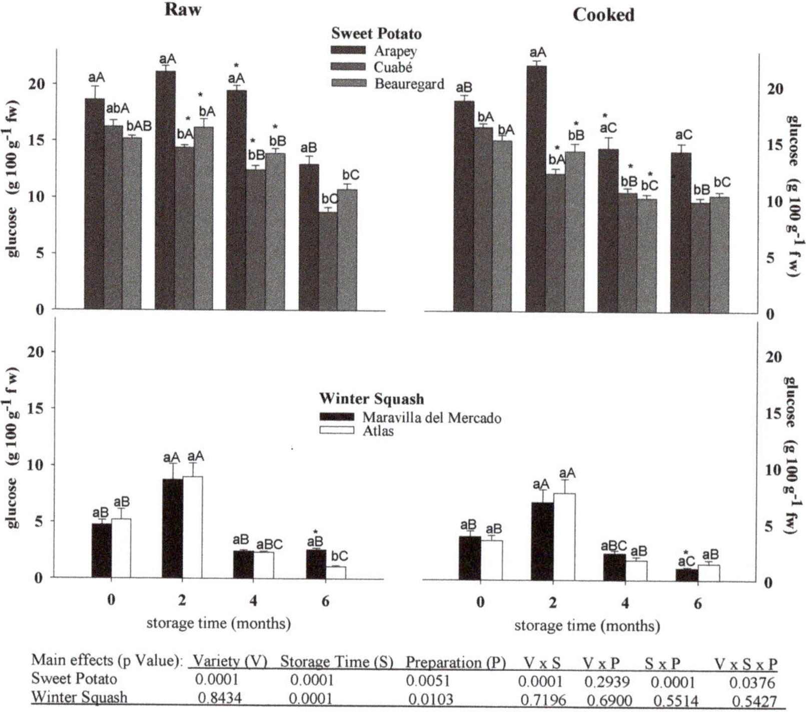

Main effects (p Value):	Variety (V)	Storage Time (S)	Preparation (P)	V x S	V x P	S x P	V x S x P
Sweet Potato	0.0001	0.0001	0.0051	0.0001	0.2939	0.0001	0.0376
Winter Squash	0.8434	0.0001	0.0103	0.7196	0.6900	0.5514	0.5427

Figure 1. Total glucose content (g 100 g^{-1} fw) in raw and cooked pulp from sweet potato and winter squash varieties stored for different times. Means ± SEM (n = 6). Different lowercase letters on each column, for sweet potatoes or winter squash, indicate statistical differences (Tukey, $p \leq 0.05$) between varieties for each process (raw or cooked) and storage time. Uppercase letters indicate statistical differences between storage times for each variety and process (raw or cooked). * denotes statistical differences by Student's t test ($p \leq 0.05$) between process (raw and cooked) in each variety and storage time.

3.2. In Vitro Bioaccessible Glucose Content in Cooked Pulp

The glucose release by in vitro digestion was higher in sweet potato (6.4 g) than winter squash cooked pulp (0.91 g). In sweet potato, the glucose released by digestion was between 8.4 and 5.1 g, without the effects of variety and storage time (Table 1).

According to these results, over time, storage caused less interference with other compounds of the pulp and probably provoked more access to enzymes in the site of action for the starch digestion. A change in the starch grains' structure could be explained most easily by the digestion of the starch [8,9,18–23,27–30].

In winter squash cooked pulp, the glucose released by in vitro digestion depended on the interaction between variety and storage time (Table 1), with strong effects of variety. Atlas cooked pulp had 42–69% less glucose released by digestion than Maravilla del Mercado during every evaluated storage time. The highest content of glucose after digestion was obtained in the second month of storage in Maravilla del Mercado cooked pulp (1.4 g) and at the end of storage in Atlas (0.8 g) (Table 1). Similar to sweet potato, the type and properties of starch are different between species and varieties and can be modified with the storage time [4,6,11,20,28], but no major changes were detected in the total amount of glucose released by digestion.

Table 1. Total glucose (g) released by in vitro digestion in 100 g cooked pulp from sweet potato and winter squash varieties stored at different times.

	Sweet Potato			Winter Squash	
Storage Time (Months)	Arapey	Beauregard	Cuabé	Maravilla del Mercado	Atlas
0	8.4 ± 1.0	5.1 ± 0.8	5.2 ± 0.8	1.1 ± 0.03 [a,B]	0.5 ± 0.07 [b,A]
2	5.6 ± 0.3	5.3 ± 0.4	6.5 ± 0.6	1.4 ± 0.06 [a,A]	0.4 ± 0.05 [b,A]
4	6.5 ± 0.6	7.5 ± 1.4	7.3 ± 0.6	1.3 ± 0.01 [a,A,B]	0.7 ± 0.07 [b,A]
6	7.7 ± 0.6	6.0 ± 0.7	6.2 ± 0.2	1.1 ± 0.09 [a,B]	0.8 ± 0.12 [a,A]

Main Effects (*p* value): Variety (V) Storage Time (S) V × S
Sweet Potato 0.1205 0.1958 0.0676
Winter Squash 0.0001 0.0542 0.0032

Means ± standard error (*n* = 6). For sweet potato or winter squash, lowercase letters on each row indicate differences (Tukey, $p \leq 0.05$) between varieties in the same storage time, and uppercase letters indicate differences between storage times for each variety. No letters indicate a lack of a statistical difference.

The percentage of glucose in vitro increased with the storage time in sweet potato (30–60%) and winter squash varieties (14–69%) (Figure 2). In sweet potato, the percentage of glucose bioaccessibility depended of the interaction of varieties and storage times. At the beginning and at the end of storage the percentage of bioaccessibility was similar between varieties, at an average of 37% at harvest and 59% at six months of storage. However, in the second month, Cuabé had twice the percentage of bioaccessible glucose than the others varieties (31%) (Figure 2). Despite the differences obtained in the total glucose content for sweet potato varieties and storage times, the cooking process was likely homogenized by the in vitro digestion. On the other hand, the winter squash pulp percentage of glucose bioaccessibility depended of the variety and storage time. It always showed a higher percentage in glucose bioaccessibility in Maravilla del Mercado than Atlas; at the same time, both varieties presented glucose bioaccessibility that rose two and three times after the second month of storage (Figure 2).

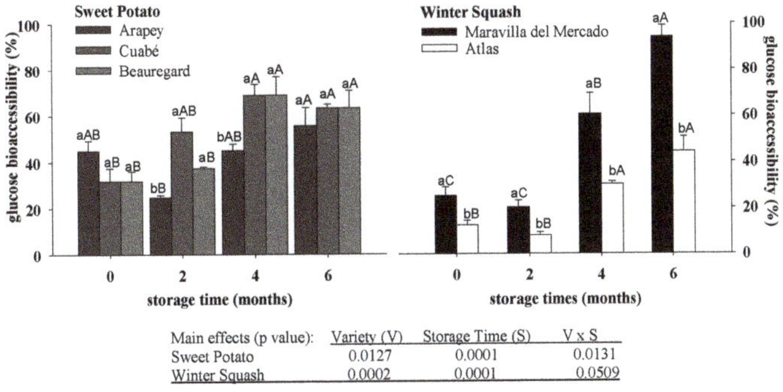

Figure 2. Percentage of in vitro bioaccessible glucose in cooked pulp from sweet potato and winter squash varieties stored for different times. Bars are means ± SEM (*n* = 6). For sweet potato or winter squash, lowercase letters on each column represent differences (Tukey $p \leq 0.05$) between varieties at each storage time, and uppercase letters represent differences between storage times for each variety.

For all varieties of the sweet potato cooked pulp, the amount of released glucose by in vitro digestion in 100 g of cooked pulp (5.1 and 8.4 g) was greater than the total glucose content in the blood of healthy adults (70–99 mg dL^{-1}, around 3.5–4.0 g of total glucose in blood) [31]. By contrast, in all cases for winter squash, cooked pulp was potentially only less than 25% of the total blood content.

4. Conclusions

The total glucose amount of sweet potato varieties was higher (14.5 g 100 g^{-1} fw) than winter squashes (4.0 100 g^{-1} fw). The Arapey sweet potato variety had the highest total content of glucose (21 g 100 g^{-1} fw) and winter squash varieties had the lowest (4 g 100 g^{-1} fw). These contents were affected by storage time, in which a prolonged storage of roots or fruits for more than two months caused a reduction in the level of total glucose. In sweet potato varieties, cooked pulp had lower glucose content than raw pulp after two months of storage time. However, in winter squash, cooking only affected the total glucose content in Maravilla del Mercado winter squash at the end of storage. The released glucose by in vitro digestion in sweet potato varieties studied here was similar and not affected by storage time, with amounts between 5.1 and 8.4 g for 100 g of cooked pulp. However, in winter squashes, Maravilla del Mercado had twice the amount of bioaccessible glucose than Atlas at every storage time (1.2 and 0.6 g 100 g^{-1} fw, respectively), and only the Maravilla del Mercado winter squash variety modified the released glucose by in vitro digestion with the storage time. Thus, the released glucose by in vitro digestion was low for both vegetables in relation to the carbohydrate daily requirement for adults (recommended dietary allowance 130 g day^{-1}) [32]. Therefore, the sweet potato varieties studied here seem to be suitable for the recommended daily carbohydrate requirement for a healthy adult, and winter squash varieties could be recommended for people with a low tolerance of glucose in blood.

Acknowledgments: The authors thank the Department of Vegetal Production of the Agronomy Faculty (UdelaR) for financial support to Pablo Latrónico for carrying outt us in the glucose determinations.

Author Contributions: Fernanda Zaccari conceived and design the research and wrote the paper. María Cristina Cabrera and Ali Saadoun assisted in the in vitro protocol for glucose determinations. All authors provide input into the interpretation of the results and approved the final manuscript.

Conflicts of Interest: The authors declare no conflict of interest.

References

1. FAO. Statistical Databases FAOSTAT. 2014. Available online: http://faostat.fao.org/es/#data/QC (accessed on 26 June 2016).
2. Lim, T.K. Cucurbita Moschata. Available online: http://link.springer.com/chapter/10.1007/978-94-007-1764-0_41#page-2 (accessed on 20 March 2016).
3. Bates, D.M.; Robinson, R.W.; Jeffrey, C. *Biology and Utilization of the Cucurbitaceae*, 1st ed.; Cornell University Press: Ithaca, NY, USA, 1990; p. 485.
4. Woolfe, J. *Sweet Potato: An Untapped Food Resource*, 1st ed.; Cambridge University Press: New York, NY, USA, 1992; p. 643.
5. Brücher, H. Carbohydrates from Roots and Tubers. In *Useful Plants of Neotropical and Their Wild Relatives*; Springer: Berlin/Heidelberg, Germany, 1989.
6. La Bonte, D.L.; Picha, D.R.; Hester, A.; Johnson, H.A. Carbohydrate-Related Changes in Sweetpotato Storage Roots during Development. *J. Am. Soc. Hortic. Sci.* **2000**, *125*, 200–204.
7. Zhang, Z.; Christopher, C.; Wheatley, C.C.; Corke, H. Biochemical changes during storage of sweet potato roots differing in dry matter content. *Postharvest Biol. Technol.* **2000**, *24*, 317–325. [CrossRef]
8. Senanayake, S.; Gunaratne, A.; Ranaweera, K.D.S.; Bamunuarachchi, A. Effect of heat-moisture treatment on digestibility of different cultivars of sweet potato (*Ipomea batatas* (L.) Lam) starch. *Food Sci. Nutr.* **2014**, *2*, 398–402. [CrossRef] [PubMed]
9. Kami, D.; Muro, T.; Sugiyama, K. Changes in starch and soluble sugar concentrations in winter squash mesocarp during storage at different temperatures. *Sci. Hortic.* **2011**, *127*, 444–446. [CrossRef]

10. K'osambo, L.M.K.; Carey, E.E.; Misra, A.K.; Wilkes, J.; Hagenimana, V. Influence of Age Farming Site and Boiling on Pro-Vitamin A content in Sweet Potato (*Ipomoea batatas* (L.) Lam) Storage Roots. *J. Food Compos. Anal.* **1998**, *11*, 305–321. [CrossRef]

11. Picha, D.H. Chilling injury, respiration and sugar changes in sweet potatoes stored at low temperature. *J. Am. Soc. Hortic. Sci.* **1987**, *112*, 497–502.

12. Smith, M.A.; Riby, L.M.; Foster, J.K. Glucose enhancement of human memory: A comprehensive research review of the glucose memory facilitation effect. *Neurosci. Biobehav. Rev.* **2011**, *35*, 770–783. [CrossRef] [PubMed]

13. Ruby, L.M.; Law, A.S.; McLaughlin, J.; Murray, J. Preliminary evidence that glucose ingestion facilitates prospective memory performance. *Nutr. Res.* **2011**, *31*, 370–377. [CrossRef] [PubMed]

14. Peters, A.; Schweiger, U.; Pellerin, L.; Hubold, L.; Oltmanns, K.M.; Conrad, M.; Schultes, B.; Born, J.; Fehm, H.L. The selfish brain: Competition for energy resources. *Neurosci. Biobehav. Rev.* **2004**, *28*, 143–180. [CrossRef] [PubMed]

15. European Food Safety Authority (EFSA). Scientific Opinion on Dietary Reference Values for Carbohydrates and Dietary Fiber. *EFSA J.* **2010**, *8*, 1–77. [CrossRef]

16. Giménez, G.; González, M.; Rodríguez, G.; Vicente, E.; Vilaró, F. Catálogo de Cultivares Hortícolas 2014. Available online: http://www.ainfo.inia.uy/digital/bitstream/item/4590/1/Catalogo-cultivares-horticolas-2014.pdf (accessed on 15 March 2015).

17. Zaccari, F.; Cabrera, M.C.; Ramos, A.; Saadoun, A. In vitro bioaccessibility of β-carotene. Ca, Mg and Zn in landrace carrots (*Daucus carota*. L.). *Food Chem.* **2015**, *166*, 365–371. [CrossRef] [PubMed]

18. Takahata, Y.; Noda, T.; Sato, T. Changes in carbohydrates and enzyme activities of sweet potato lines during storage. *J. Agric. Food. Chem.* **1995**, *43*, 1923–1929. [CrossRef]

19. Corrigan, V.K.; Irving, D.E.; Potter, J.F. Sugars and sweetness in buttercup squash. *Food Qual. Preference* **2000**, *11*, 313–322. [CrossRef]

20. Picha, D.H. Carbohydrates changes in sweet potatoes during curing and storage. *J. Am. Soc. Hortic. Sci.* **1987**, *111*, 89–92.

21. Jiang, Y.; Chan, C.; Tao, X.; Wang, I.; Zhang, Y. A proteomic analysis of storage stress responses in *Ipomoea batatas* (L.) Lam. Tuberous root. *Mol. Biol. Rep.* **2012**, *39*, 8015–8025. [CrossRef] [PubMed]

22. Shekhar, S.; Mishra, D.; Buraohain, A.K.; Charkraborty, S.; Charkaborty, N. Comparative analysis of phytochemicals and nutrient availability in two contrasting cultivars of sweet potato (*Ipomoea batatas* L.). *Food Chem.* **2015**, *173*, 957–965. [CrossRef] [PubMed]

23. Zhang, T.; Oates, C.G. Relationship between α-amylase degradation and physicochemical properties of sweet potato starches. *Food Chem.* **1999**, *65*, 157–163. [CrossRef]

24. Stevenson, D.G.; Yoo, S.-H.; Hurst, P.L.; Jane, J.-L. Structural and physicochemical characteristics of winter squash (*Cucurbita maxima* D.) fruit starches at harvest. *Carbohydr. Polym.* **2005**, *59*, 153–163. [CrossRef]

25. McGillivray, J.H. Soluble solids content of different regions of watermelons. *Plant Physiol.* **1947**, *22*, 637–640. [CrossRef]

26. Scott, G.W.; MacGillivray, J.H. Variation solids of the juice from different regions in melon fruits. *Hilgardia* **1940**, *13*, 69–79. [CrossRef]

27. Trancoso-Reyes, N.; Ochoa-Martinez, L.A.; Bello-Perez, L.A.; Morales-Castro, L.; Estevez-Santiago, R.; Olmedilla-Alonso, B. Effect of pre-treatment on physicochemical and structural properties, and the bioaccessibility of beta-carotene in sweet potato flour. *Food Chem.* **2016**, *200*, 199–205. [CrossRef] [PubMed]

28. Aina, A.J.; Falade, K.O.; Akingbala, J.O.; Titus, P. Physicochemical Properties of Caribbean Sweet Potato (*Ipomoea batatas* (L.) Lam) Starches. *Food Bioprocess Technol.* **2012**, *5*, 576–583. [CrossRef]

29. Lai, Y.C.; Huang, C.L.; Chan, C.F.; Lien, C.Y.; Liao, W.C. Studies of sugar composition and starch morphology of baked sweet potatoes (*Ipomoea batatas* (L.) Lam). *J. Food Sci. Technol.* **2013**, *50*, 1193–1199. [CrossRef] [PubMed]

30. Mennah-Govela, Y.A.; Bornhorst, G.M. Acid and moisture uptake in steamed and bolied sweet potatoes and associated structure changes during in vitro gastric digestion. *Food Res. Int.* **2016**, *88*, 247–255. [CrossRef]

31. Moghissi, E.S.; Mary, T.; Korytkowski, M.T.; DiNardo, M.; Einhorn, D.; Hellman, R.; Hirsch, I.B.; Inzucchi, S.E.; Ismail-Beigi, F.; Kirkman, M.S.; et al. American Association of Clinical Endocrinologists nad American Diabetes Association consensus statement on in patient glycemic control. *Endocr. Pract.* **2009**, *15*, 353–369. [CrossRef] [PubMed]

32. Institute of Medicine (IOM). *Dietary Reference Intakes for Energy, Carbohydrate, Fiber, Fat, Fatty Acids, Cholesterol, Protein, and Amino Acids*; National Academies Press: Washington, DC, USA, 2001; p. 1319.

Communication

Unequivocal Identification of 1-Phenylethyl Acetate in Clove Buds (*syzygium aromaticum* (L.) Merr. & L.M. Perry) and Clove Essential Oil

Klaus Gassenmeier [1,*], **Hugo Schwager** [1], **Eric Houben** [2] and **Robin Clery** [1]

[1]   Givaudan (Schweiz) AG, Dübendorf 8600, Switzerland; Hugo.Schwager@givaudan.com (H.S.);
      Robin.Clery@givaudan.com (R.C.)
[2]   Givaudan Nederland BV, GP Naarden 1411, Netherlands; Eric.Houben@givaudan.com
*   Correspondence: klaus.gassenmeier@givaudan.com; Tel.: +41-44-824-2521

Academic Editors: Theo H. Varzakas and Charalampos Proestos
Received: 5 May 2017; Accepted: 21 June 2017; Published: 27 June 2017

Abstract: The natural occurrence of 1-phenylethyl acetate (styrallyl acetate) was confirmed in commercially available dried clove buds and also in the hydrodistilled oil from clove buds. This confirms previous reports and other anecdotal evidence for its occurrence in nature.

Keywords: 1-phenylethyl acetate; styrallyl acetate; alpha methyl benzyl acetate; clove; *Syzygium aromaticum*; volatile composition

---

1. Introduction

1-Phenylethyl acetate (the Federal Emergency Management Agency (FEMA): 2684, CAS 93-92-5) is a well-known flavor constituent admitted in many countries as a flavor ingredient. Its flavor has been described as sweet and fruity, tropical, mango, woody, musty, honey like with floral powdery nuances [1]. It is available in artificial and natural versions from various suppliers, and may also be known as styrallyl acetate or alpha methyl benzyl acetate.

According to the Volatile Compounds in Food database [2], its occurrence in food is reported in avocado [3], honey [4], melon [5] and strawberry guava [6]. Other publications state its occurrence in pineapple [7] and banana [8]. Nevertheless, none of the cited publications provides sufficient evidence for the occurrence of the title compound in food, especially when the criteria of the International Organisation of the Flavour Industry—Working Group of Methods and Analysis (IOFI WGMA) recommendation [9] are considered, which require use of an authentic standard and two independent methods of identification, e.g., mass spectrum and retention index. The occurrence of 1-phenylethyl acetate (1-PA) is mentioned in steam-distilled and expressed clove bud oil by Brian Lawrence [10] and in several Cinnamonum species [11], but no analytical details are given in those publications. This has led to remaining doubts about its occurrence in nature and some analytical service labs would regard 1-PA as not yet having been identified in nature. Consequently, flavors containing 1-PA would be regarded as not natural, regardless of whether 1-PA had been obtained from a natural source or not.

Since we had prior evidence in-house of the occurrence of 1-PA in a clove essential oil (un-published results), we re-investigated clove bud concerning the occurrence of the title compound following the recommendation in [9].

2. Materials and Methods

Clove buds were obtained from a local retail store. Sample A: Clove buds, whole (Migros, Dübendorf, Switzerland). The exact country of origin could not be given by the supplier, but would be Madagascar, Indonesia, Sri Lanka or Comores (communication Remo Kessler, Customer service, M-Infoline, Zurich, Switzerland). Sample B: Clove buds, whole, organic (Coop, Volketswil, Swithzerland), Origin: India and/or Sri Lanka. Sample C: Clove buds, ground (Coop, Volketswil, Swithzerland), Origin: Madagascar.

1-Phenylethyl acetate, eugenol, eugenol acetate and beta-caryophyllene were obtained from Aldrich, Switzerland.

Essential oil isolation and analysis: Clove buds (Sample A; 55 g) were finely ground, mixed with water (250 mL), placed in a round bottom flask (500 mL) and hydrodistilled using a Clevenger type apparatus. Distillation was conducted for 1 h and the oil collected in an oil receiver. The essential oil was recovered from the condenser. A dilution 1:5 in methyl tert.butyl ether was injected in the gas chromatography (GC) with a split ratio of 1:20.

Solid Phase Micro Extraction (SPME): 0.5 g of clove buds were finely ground and sealed in a 20 mL headspace vial. The headspace in the vial was extracted by automated SPME using a CTC autosampler (Agilent, Santa Clara, CA, USA). Equilibration at 40 °C for 10 min, then exposure of the SPME fiber (50/30 μm DCB/CAR/PDMA, Supelco; Bellafonte, PA, USA) for 10 min. The fiber was desorbed directly in a split/splitless inlet (230 °C) for 30 s in splitless mode.

Gas Chromatography Mass Spectrometry: An Agilent Gas Chromatograph 6890 coupled to mass spectrum detector (MSD) 5975B and flame ionization detection (FID) was used. The effluent was split 1:1 between FID and mass spectrum (MS). Linear retention index was calculated on an alkane scale. Column: CP-WAX 60 m × 0.32 mm ID × 0.25 μm film thickness (Agilent, Santa Clara, CA, USA). Temperature program: from 35 °C, held for 2 min, to 250 °C at 4 °C/min. Carrier gas: He, linear velocity: 25 cm/s. Injection temperature: 250 °C. Injection volume: 1 μL. Injection mode: split (20:1). FID (250 °C). MS Interface temperature: 280 °C; MS mode: EI at 70 eV; Ion source temperature: 150 °C; Mass range 29–250 u. Data handling was carried out using MSD Chemstation (Agilent, Santa Clara, CA, USA).

Chiral separation was conducted using a β-DEX 120 column (Sigma-Aldrich, Saint Louis, MO, USA, 30 m × 0.25 mm × 0.25 μm film thickness, Nr. 24304a). Carrier gas was helium at 25 cm/s. Temperature program: from 60–130 °C at 2 °C/min then raised to 200 °C at 40 °C/min. The enatiomeric excess was determined by integrating *m/z* 104 and *m/z* 105. For both isomers, clean mass spectra were obtained after background subtraction.

Identification: Mass spectra and retention index were compared with reference samples.

3. Results

Analysis of the obtained hydrodistilled oil (Sample A) revealed Eugenol (86.3%), beta-Caryophyllene (6.3%) and Eugenyl acetate (4.1%) as the main constituents. For the identification of the 1-PA the relevant retention time window was screened for the occurrence of the most typical mass fragments *m/z* 104, 122 and 164. At a retention time of 31.26 min, the mass spectrum in Figure 1 (top) was obtained after background subtraction. It matched the mass spectrum of the reference (Figure 1, middle). The measured retention index was 1697 compared to 1694 of the authentic reference. Based on the FID signal integration, the area percent of 1-PA was estimated to be around 0.05%. This was a significantly higher value compared to a previous in-house analysis, where ground clove bud was extracted and 1-PA showed a concentration of 0.0027% (internal communication). Lawrence [10] reported the sum amount of 1-PA and two other co-eluting constituents in distilled clove, and in expressed oil from Madagascar at 0.1% and 0.21%, respectively, which closely matches our measured value of 0.05%.

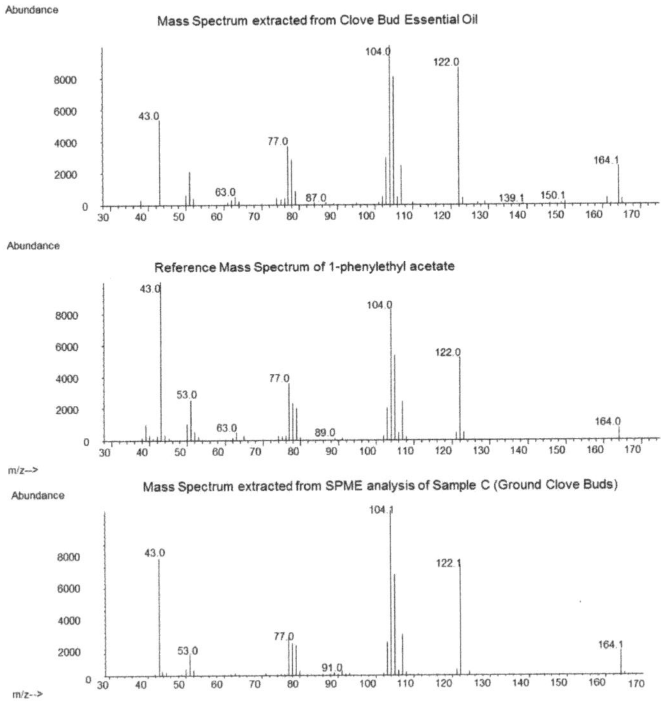

Figure 1. Comparison of the mass spectrum obtained from clove bud essential oil (top), reference mass spectrum from 1-phenylethyl acetate (1-PA) (middle) and solid-phase microextraction (SPME) of ground clove buds (sample C, bottom).

The heating during distillation may introduce chemical changes to a product, and essential oils may contain compounds not present in the native plant. To exclude this possibility, we conducted a SPME extraction of clove bud samples A, B and C. In those three samples, we could also detect the mass spectrum of 1-PA at the correct retention index (mass spectrum of 1-PA from SPME of sample C see Figure 1, bottom). The highest intensity was found in sample C (origin Madagascar), but no quantification was conducted.

The formation pathway of 1-phenylethyl acetate is unclear; there could be an enzymatic or chemical process. The enzymatic processes of biosynthesis often lead to products with an enantiomeric excess of one stereo isomer. To elucidate this, we conducted a chiral analysis of the 1-phenylethyl acetate. Peak identification was based on mass spectrometry and comparison to a sample of 1-PA racemate. In the essential oil obtained from clove buds, we found a significant enantiomeric excess of 62% for the first eluting enantiomer. Lacking enantiomeric pure 1-PA as reference for either of the two 1-PA enantiomers, we could not assign the configuration (R or S) of the major isomer.

4. Discussion and Conclusions

1-Phenyethyl acetate was unequivocally identified in clove bud essential oil and in ground clove buds. In lab-distilled essential oil obtained from clove buds, we found an enantiomeric excess of 62% of the first eluting enantiomer. On the basis of these observations, we conclude that 1-phenylethyl acetate occurs naturally in clove buds and is probably the product of enzymatic biosynthesis, and not an artifact of subsequent processing.

Author Contributions: K.G. and H.S. conceived and designed the experiments; H.S. performed the experiments related to extraction and sample preparation; E.H. designed and executed the chiral separation; K.G. and R.C. wrote the paper.

Conflicts of Interest: The authors declare no conflict of interest.

References

1. Mosciano, G. Organoleptic characteristics of flavor materials. *Perfum. Flavorist* **2006**, *31*, 46–49.
2. Volatile Compounds in Food 16.3. Available online: http://www.vcf-online.nl/VcfHome.cfm (accessed on 8 March 2017).
3. Yamaguchi, K.; Nishimura, O.; Toda, H.; Mihara, S.; Shibamoto, T. Chemical studies on tropical fruits. In *Instrumental Analysis of Foods: Recent Progress*; Charalambous, G., Inglett, G., Eds.; Academic Press: San Diego, CA, USA, 1983; pp. 93–117.
4. Plutowska, B.; Chmiel, T.; Dymerski, T.; Wardencki, W. A headspace solid-phase microextraction method development and its application in the determination of volatiles in honeys by gas chromatography. *Food Chem.* **2011**, *126*, 1288–1298. [CrossRef]
5. Chaparro-Torres, L.A.; Bueso, M.C.; Fernández-Trujillo, J.P. Aroma volatiles obtained at harvest by HS-SPME/GC-MS and INDEX/MS-E-nose fingerprint discriminate climacteric behaviour in melon fruit. *J. Sci. Food Agric.* **2016**, *96*, 2352–2365. [CrossRef] [PubMed]
6. Pino, J.A.; Marbot, R.; Vázquez, C. Characterization of volatiles in strawberry guava (*Psidium cattleianum* Sabine) fruit. *J. Agric. Food Chem.* **2001**, *49*, 5883–5887. [CrossRef] [PubMed]
7. Steingass, C.B.; Carle, C.; Schmarr, H.G. Ripening-dependent metabolic changes in the volatiles of pineapple (*Ananas comosus* (L.) Merr.) fruit: I. Characterization of pineapple aroma compounds by comprehensive two-dimensional gas chromatography-mass spectrometry. *Anal. Bioanal. Chem.* **2015**, *407*, 2591–2608. [CrossRef] [PubMed]
8. Pino, J.A.; Febles, Y. Odour-active compounds in banana fruit cv. Giant Cavendish. *Food Chem.* **2013**, *141*, 795–801. [CrossRef] [PubMed]
9. Brevard, H.; Cachet, T.; Chaintreau, A.; Demyttenaere, J.C.R.; Dijkhuizen, A.; Gassenmeier, K.; Joulain, D.; Koenig, T.; Leijs, H.; Liddle, P.; et al. Statement on the identification in nature of flavouring substances, made by the Working Group on Methods of Analysis of the International Organization of the Flavour Industry (IOFI). *Flavour Fragr. J.* **2006**, *21*, 185.
10. Lawrence, B.M. Major tropical spices—Clove. In *Essential Oils 1976–1977*; Allured Publ Corp.: Carol Stream, IL, USA, 1979; pp. 84–145.
11. Cheng, B.Q.; Yu, X.J.; Ding, J.K.; Xu, Y.; Sun, H.D.; Ma, X.X.; Yi, Y. *Cinnamomum Plant Resources and Their Aromatic Constituents*; Yunnan Science and Technology Press: Yunnan, China, 2001.

Article

Influence of Food Characteristics and Food Additives on the Antimicrobial Effect of Garlic and Oregano Essential Oils

Juan García-Díez [1,*], Joana Alheiro [1], Ana Luísa Pinto [1], Luciana Soares [1], Virgílio Falco [2], Maria João Fraqueza [3] and Luís Patarata [1]

[1]   CECAV, Centro de Ciência Animal e Veterinária, Universidade de Trás-os-Montes e Alto Douro, Quinta de Prados, Vila Real 5000-801, Portugal; jo_alheiro@hotmail.com (J.A.); pinto_aluisa@hotmail.com (A.L.P.); lucimlsoares@hotmail.com (L.S.); lpatarat@utad.pt (L.P.)

[2]   CQ-VR, Centro de Química-Vila Real (CQ-VR), Universidade de Trás-os-Montes e Alto Douro, Quinta de Prados, Vila Real 5000-801, Portugal; vfalco@utad.pt

[3]   CIISA, Faculty of Veterinary Medicine, University of Lisbon, Avenida da Universidade Técnica, Pólo Universitário do Alto da Ajuda, Lisbon 1300-477, Portugal; mjoaofraqueza@fmv.ulisboa.pt

*   Correspondence: juangarciadiez@gmail.com; Tel.: +351-259350000; Fax: +351-259350480

Academic Editors: Theo H Varzakas and Charalampos Proestos

Received: 6 May 2017; Accepted: 8 June 2017; Published: 10 June 2017

Abstract: Utilization of essential oils (EOs) as antimicrobial agents against foodborne disease has gained importance, for their use as natural preservatives. Since potential interactions between EOs and food characteristics may affect their antimicrobial properties, the present work studies the influence of fat, protein, pH, a_w and food additives on the antimicrobial effect of oregano and garlic EOs against *Salmonella* spp. and *Listeria monocytogenes*. Results showed that protein, pH, a_w, presence of beef extract, sodium lactate and nitrates did not influence their antimicrobial effect. In contrast, the presence of pork fat had a negative effect against both EOs associated with their dilution of the lipid content. The addition of food phosphates also exerts a negative effect against EOs probably associated with their emulsification properties as observed with the addition of fat. The results may help the food industry to select more appropriate challenges to guarantee the food safety of foodstuffs.

Keywords: essential oil; *Salmonella* spp.; *Listeria monocytogenes*; food additives; food composition

---

1. Introduction

Use of essential oils (EOs) as flavoring agents by the food industry is common. However, their antimicrobial properties against foodborne pathogens have increased their interest as a source of natural preservatives [1,2]. In dry-cured meat products, utilization of EOs has gained interest in the potential control of pathogens. Considering the strong aromatization of these sausages, resulting from the smoking process and seasonings [3,4], EOs as antimicrobial agents could be used since it is expected that the sensory impact of EOs will be mitigated by the global aroma of the product.

The antimicrobial effect of EOs is different in foodstuffs than in in vitro studies [5]. Since factors such as pH, a_w, food composition or potential interactions with food additives influence the antimicrobial effect of the EOs [6], a previous in vitro screening against specific foodborne pathogens tested in specific food-like growth media could be an important approach to highlight potential interactions between EOs and food characteristics. Thus, the present work is aimed at studying the influence of fat and protein levels, different pH and a_w and the presence of sodium nitrite, commercial phosphates and sodium lactate on the inhibitory properties of oregano and garlic EOs against *Salmonella* spp. and *L. monocytogenes* isolated from meat products.

2. Material and Methods

2.1. Gas Chromatography-Mass Spectrometry Analysis of Essential Oils

EOs of the spices of garlic (*Allium sativum*, bulbs) and oregano (*Origanum vulgare*, leaves), commonly used to manufacture dry-cured meat products, as reported by Melo et al. [7], were selected. All EOs and their technical characteristics were kindly provided by Ventós Chemicals (Barcelona, Spain). The Gas chromatography-mass spectrometry analysis was carried out as described elsewhere [8].

2.2. Microorganisms and Growth Conditions

Stock cultures of *Salmonella* spp. and *L. monocytogenes* (Table 1) isolated either from traditional dry-cured fermented sausages during their manufacturing or from the environment of their production were identified by a species-specific PCR technique [9]. Each microorganism was maintained at −18 °C and subcultured twice in brain heart infusion (BHI-Biokar, Beauvais, France). Incubation for *Salmonella* spp. was done at 37 °C, while *L. monocytogenes* was incubated at 30 °C. Overnight cultures in BHI were streaked in BHI agar and incubated during 18 to 24 h. To prepare the inoculum for the sensitivity test to EOs, a suspension of isolated colonies in BHI agar was made in NaCl 0.85%. The turbidity of the suspension was adjusted to 0.5 McFarland standard (Biomerieux, Marcy-l'Etoile, France).

Table 1. Strains used in the experiment.

Microorganisms	Strain	Source [1]
Salmonella spp.	MPI-B-S07	Chouriço batter
	EDS-E-S02	Environment of meat products preparation
L. monocytogenes	EDS-B-LM02	Chouriço batter
	MPI-E-LM01	Environment of meat products preparation

[1] Strains isolated from meat products or the environment of its production are from our laboratory collection.

2.3. Antimicrobial Effect on Disk Diffusion Assay

The antimicrobial effect of EOs was screened by the disk diffusion assay (DDA) as described by Zaika [10], but with some modifications. Petri plates prepared with 20 mL of Mueller–Hinton agar (MHA, Biokar. Beauvais, France) were dried, and 100 μL of standardized inoculum suspension (ca. 8 log CFU/mL) were poured and uniformly spread. Filter paper disks (Whatman No. 1, 6-mm diameter, GE Healthcare, Madison, WI, USA) containing 20 μL of each EOs were applied to the surface of the previously seeded agar plates of MHA. The plates were kept at 4 °C for 2 h to allow dispersion and were incubated overnight at the optimum growth temperature of each microorganism under study (as above mentioned) during 18 to 24 h. The antimicrobial activity was visually evaluated as the inhibition zone surrounding the disk, and their diameters, including the disk diameters, were measured in mm. The results representing the net zone of inhibition including the diameter (6 mm) of the paper disk are the mean of 3 determinations for each isolate tested.

2.4. Determination of MIC and MBC

The minimum inhibitory concentration (MIC) and minimum bactericidal concentration (MBC) were studied for both EOs. The dilutions of the EOs were established based on the inhibitory profile with the DDA. The assay was based on the procedure of the Clinical and Laboratory Standards Institute [11] with 96-well microtiter plates. The MIC was considered the lowest concentration of EOs at which bacteria failed to grow, as detected by the unaided eye, matching with the negative control without inoculation included in the test. The visual evaluation was complemented with the seeding of a 10-μL loop in MHA to confirm the absence of growth. To evaluate the MBC, 10 μL of each well, in which no microbial growth was observed, were spread into MHA plates and incubated for 24 h. The MBC was considered as the lowest concentration determining a reduction in the population of 99.9%.

2.5. Preparation of Food Model Media

Influence of Fat, Protein, pH and a_w

The effect of different levels of fat and protein, pH and a_w on the antimicrobial effect of EOs was performed using a food model media. *L. monocytogenes* or *Salmonella* spp. were used as indicator strains. EOs of oregano and garlic were used at 0.05% and 0.005%, respectively. The influence of 2.5%, 5% and 10% of fat was studied by the addition of pork fat, purchased at a local supermarket and previously sterilized by autoclaving before being added to Mueller-Hinton broth supplemented with 1% agar (MHB, Biokar. Beauvais, France). The fat was emulsified with the broth with an Ultra-Turrax (IKA, Staufen, Germany). The influence of protein was tested by the addition of beef extract (Oxoid, Hampshire, UK) at 10%, 15% and 20% to MHB. The influence of pH was tested by the addition of lactic acid (Panreac Applichem, Barcelona, Spain) to MHB to achieve a final pH of 4.5, 5.5 or 6.5 (Crison, Barcelona, Spain) with a penetration probe (Mettler-Toledo, Giesen, Germany). The influence of a_w was tested by adjusting the MHB to a final a_w of 0.91, 0.94, 0.97 by the addition of NaCl as described by Troller et al. [12]. The a_w was measured in a Hygroscope DT apparatus (Rotronic, Zurich, Switzerland) with a WA40 cell maintained at 20 ± 2 °C. All of the levels of fat and protein concentration, pH and a_w were selected to simulate the physical and chemical conditions of a traditional dry-cured meat product.

2.6. Influence of Food Additives (Sodium Nitrite, Commercial Phosphates and Sodium Lactate)

The influence of food additives on the antimicrobial effect of EOs of oregano and garlic was assessed by the addition of nitrites, phosphates and sodium lactate to MHB as follows: 150 ppm sodium nitrite (Merck, Darmstadt, Germany), 0.5% commercial phosphate (E451 plus E452, BK Giulini, Mannhein, Germany) and 3.3% sodium lactate (Sigma-Aldrich, St. Luis, MO, USA).

2.7. Microbial Preparation

Two strains of *Salmonella* spp. and two of *L. monocytogenes*, identified by a species-specific PCR technique as indicated above [9], were used in the experiment. The strains used (Table 1) did not present sensibility differences against the antimicrobial effect of EOs of oregano and garlic as reported elsewhere [8]. Single strain cultures of each pathogen were inoculated, in duplicate, in test tubes with 10 mL of culture medium prepared with the specific modification. The inoculation was made to achieve an initial contamination of about 5.7 log CFU/mL. Inoculated tubes were incubated at 37 °C for *Salmonella* spp. or 30 °C for *L. monocytogenes*. Counts were performed after 4, 8, 12, 24 and 36 h of incubation by serial ten-fold dilution prepared from 1 mL of the culture in xylose lysine desoxycholateagar (Oxoid, Hampshire, UK) for *Salmonella* spp. and Oxford agar (Oxoid, Hampshire, UK) for *L. monocytogenes*. Tests with food additives were performed only until 24 h of incubation. The experiment was carried out in triplicate, and results are expressed as the log CFU/mL of culture medium.

2.8. Statistical Analysis

The influence of fat and protein levels, a_w, pH and food additives on the antimicrobial effect of oregano and garlic EOs against each pathogen was carried out by analysis of variance (ANOVA), evaluating the combined effect of the presence of EOs and the level of the composition modification studied, for each incubation time. The Tukey–Kramer test was used to determine the significant differences ($p < 0.05$) among means. Statistical analysis was done with SPSS 19.0 software (SPSS Inc., Chicago, IL, USA) for Windows 8 (Redmont, Washington, DC, USA), considering $p < 0.05$ as statistically significant.

3. Results and Discussion

3.1. Chemical Composition and Antimicrobial Properties of EOs

Information about the utilization of EOs to improve both the food safety and shelf-life of foodstuffs is scarce and mainly aimed at fresh foodstuffs [13–15], although some reports studied its application in meat products [16,17]. Since meat products presented specific characteristics as variable protein and fat levels, pH, a_w or additives, the influence of these characteristics on the inhibitory effect of EOs should be previously assessed in vitro to address their further potential application in meat products. Although some works studied the influence of food composition on the antimicrobial effect of EOs in specific foodstuffs [5], the present work is, to the best knowledge of the authors, the first report that studied the influence of food composition and food additives commonly present in meat products.

EOs of garlic and oregano displayed a noticeable inhibitory activity against *Salmonella* spp. and *L. monocytogenes* [18,19] based on their main chemical compounds, thymol and sulfur compounds, respectively [1] (Tables 2 and 3). Regarding the MIC and the MBC (Table 3), results were in accordance with DDA. The large inhibition halos observed for *L. monocytogenes* for EOs of garlic and oregano were in accordance with the low MIC and MBC values. In contrast, the higher MIC and MBC of garlic essential oil for *Salmonella* spp. were in accordance with the lower halo size.

Table 2. Chemical composition of essential oils of oregano and garlic determined by GC-MS.

Garlic essential oil		Oregano essential oil	
Compounds	**%**	**Compounds**	**%**
Diallyl sulfide	8.36	á-Pinene	0.27
Methyl allyl disulfide	2.76	á-Terpinolene	0.27
diallyl disulfide	18.86	*p*-cymene	0.99
Methyl allyltrisulfide	9.04	á-Terpinene	1.29
1,3,5 trithiane	0.75	Linalool	0.21
2-vinil-1,3-dithiane	0.75	Thymol	93.34
diallyltrisulfide	33.82	*Trans*-Caryophyllene	0.72
Hexamethylenesulfoxide	0.24	Germacrene	0.13
Methyl allyl disulfide	2.75		
Diallyltetrasulfide	10.97		

Table 3. Zones of growth inhibition (mm; mean ± standard deviation) with the DDA (disk diffusion assay), minimal inhibitory concentration (MIC) and minimal bactericide concentration (MBC) of garlic and oregano essential oils against *Salmonella* spp. and *L. monocytogenes*.

Essential Oil	Assay	*Listeria monocytogenes*	*Salmonella* spp.
garlic	DDA (mm)	10.5 ± 1.6	15.03 ± 2.6
	MIC (%)	2	0.0125
	MBC (%)	4	2
oregano	DDA (mm)	46.5 ± 3.2	36.4 ± 1.3
	MIC (%)	0.005	0.005
	MBC (%)	>0.005	>0.005

3.2. Effect of Fat, Protein, a_w, pH and Food Additives (Sodium Nitrite, Sodium Lactate and Food Phosphates)

Regarding the influence of the fat level (Figure 1), the inhibitory effect of EOs of garlic and oregano decreased as the fat level increases. Oregano EOs displayed a noticeable inhibitory effect at a 2.5% and 5% fat level after 4 and 8 h. of incubation. However, 12 h later, counts of *Salmonella* spp. and *L. monocytogenes* were similar in samples with garlic EOs and control. Oregano EOs showed higher antimicrobial activity than garlic EOs ($p < 0.05$), although no statistical differences were observed between samples with garlic EOs and control ($p > 0.05$). A high level of fat exerted a negative effect of the antimicrobial effect of oregano EOs in accordance as reported in the literature [6]. In addition, the antimicrobial effect of oregano essential oil, in the presence of fat, was similar for *Salmonella* spp.

and *L. monocytogenes*, contrary to what was reported by Solórzano-Santos and Miranda-Novales [20]. The decrease in the inhibitory activity at high fat levels could be explained by its protective effect or by the dilution effect of EOs in the fat, decreasing the contact between EOs and pathogens [21]. Our results are in accordance with Smith-Palmer et al. [19], who reported a higher inhibition of *L. monocytogenes* and *S. enteritidis* in low-fat soft cheese than in high-fat soft cheese. In addition, the increase of the microbial counts along the cheese ripening is also observed in the current study. The negative effect of fat was also reported by Cava et al. [22], who observed a decrease in the antimicrobial activity of cinnamon and clove EOs against *L. monocytogenes* in whole milk compared to skimmed milk. Moreover, Singh et al. [23] observed similar results against *L. monocytogenes* in low-fat hotdog. The high inhibitory effect observed at the lowest fat level could be achievable by the high contact between EOs and foodborne pathogens after 4 h. However, the decrease in the antimicrobial effect along the study suggests a dispersion of the EOs on the lipid content of the food model media. It suggests that the decrease of the inhibitory effect is caused by a dilution effect. In the case of the EOs of garlic, the reduction of its antimicrobial properties could be associated with the degradation of sulfur compounds by the chemical reactions of lipid oxidation, as suggested by Druum et al. [24].

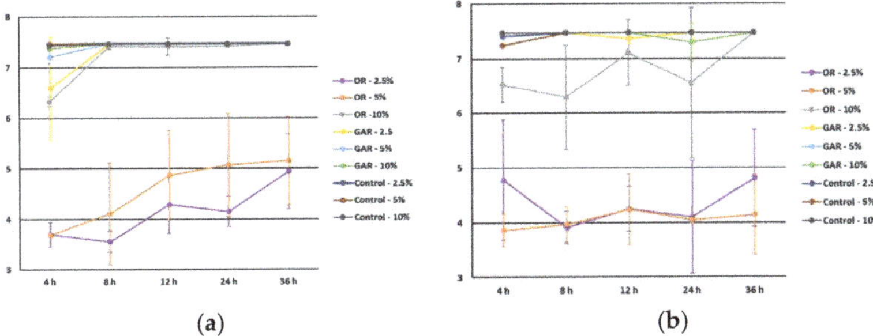

Figure 1. Influence of fat level on the antimicrobial effect of oregano (OR) and garlic (GAR) essential oils against *Salmonella* spp. (**a**) and *Listeria monocytogenes* (**b**).

Protein level (Figure 2) did not influence ($p > 0.05$) the antimicrobial effect of oregano and garlic EOs although lower microbial counts were observed at the higher level tested. Other studies showed a reduction of the antimicrobial effect of thyme and oregano EOs in the presence of beef extract [6], as well as in the presence of minced fish [25] against *L. monocytogenes*. This suggests that the presence of high levels of protein could decrease the interaction between EOs and microorganisms due to the formation of a three-dimensional matrix of proteins that acts as a barrier [26,27] or by their hydrophobic properties [28,29]. Thus, the difficult distribution of the EOs in the food model media may explain the scarce antimicrobial effect.

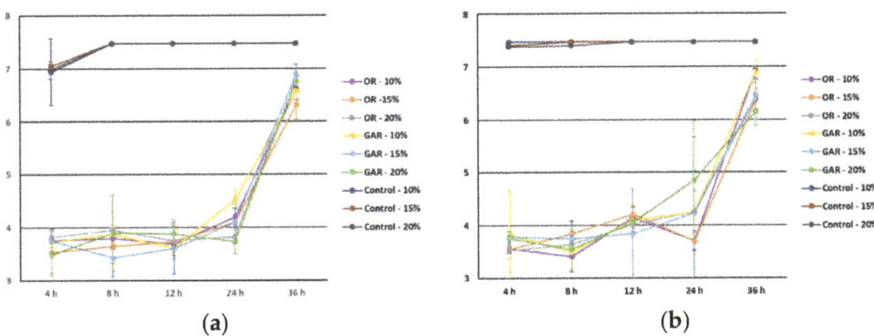

Figure 2. Influence of protein level (Prot) on the antimicrobial effect of oregano (OR) and garlic (GAR) essential oils against *Salmonella* spp. (**a**) and *Listeria monocytogenes* (**b**).

The antimicrobial effect of EOs against *Salmonella* spp. and *L. monocytogenes* was noticeable at the three levels of a_w tested and lower than control samples ($p < 0.001$), although the results were similar in samples with EOs ($p > 0.05$).

The influence of a_w (Figure 3) on the antimicrobial effect of EOs was slightly higher at reduced a_w, particularly at 0.91. The a_w range for growth of *Salmonella* spp. and *L. monocytogenes* is 0.94 to 0.99 and 0.92 to 0.99, respectively [29,30]. Since *L. monocytogenes* is a Gram-positive bacteria, a higher susceptibility to the antimicrobial effect of Eos was expected due to the absence of the outer membrane that acts as a protective barrier. However, its capacity to adapt to osmotic stress [31] may explain the higher counts compared to *Salmonella* spp. at a_w 0.91.

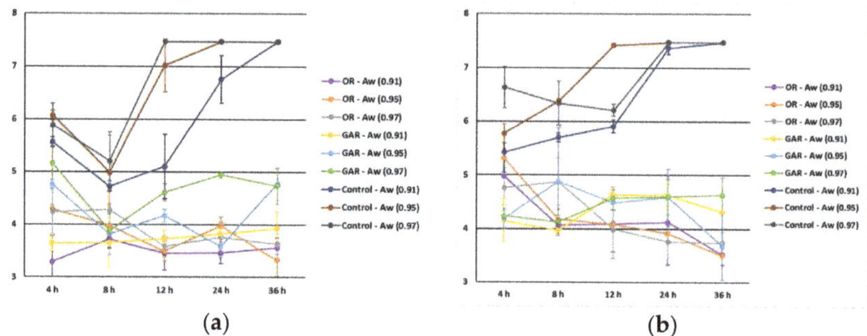

Figure 3. Influence of a_w level on the antimicrobial effect of oregano (OR) and garlic (GAR) essential oils against *Salmonella* spp. (**a**) and *Listeria monocytogenes* (**b**).

The influence of pH (Figure 4) on the antimicrobial effect of EOs was similar as observed for a_w. However, no statistical differences were observed among the three pH levels tested ($p > 0.05$). Lower counts of *Salmonella* spp. and *L. monocytogenes* were observed at pH 4.5 in accordance with Gutierrez et al. [6]. These decreases could be associated with an increase in the hydrophobicity of EOs at low pH that facilitates the dissolution of the lipids presented in the outer membrane of foodborne pathogens [32,33].

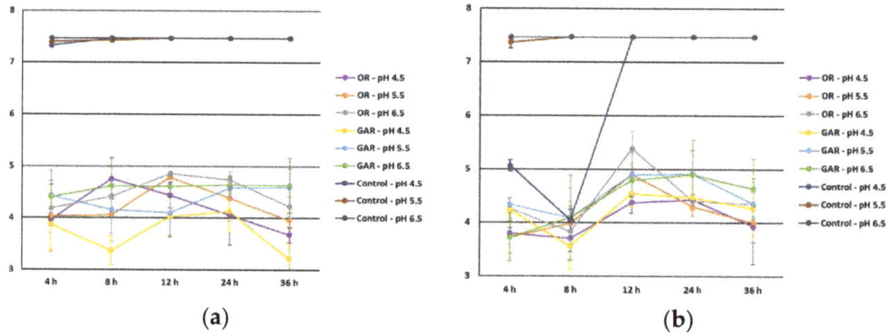

Figure 4. Influence of pH level on the antimicrobial effect of oregano (OR) and garlic (GAR) essential oils against *Salmonella* spp. (**a**) and *Listeria monocytogenes* (**b**).

Regarding food additives, the inhibitory effect of EOs was observed in the presence and absence of sodium nitrite (Figure 5) and sodium lactate (Figure 6). This indicates that a decrease in the microbial counts could not be attributed to a synergic effect with these additives ($p > 0.05$). Sodium nitrite is mainly used as a preservative, to control microbial development, especially *Clostridium botulinum*, although its addition enhances some characteristics of foodstuffs, such as the aroma and typical color of cured meat products, by the formation of nitrosylmyoglobin [34]. Although some works [35] suggested an antimicrobial effect of nitrites against Enterobacteriaceae in dry-cured chorizo, this effect should be carefully interpreted since other factors, such as the decrease in the pH or a_w values, with special relevance in this kind of product, may directly affect the survival of foodborne pathogens.

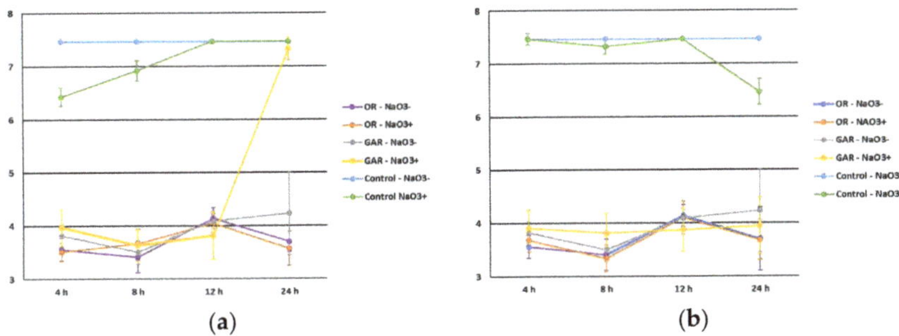

Figure 5. Influence of sodium nitrite (absence, NaNO3$^-$; presence, NaNO3$^+$) level on the antimicrobial effect of oregano (OR) and garlic (GAR) essential oils against *Salmonella* spp. (**a**) and *Listeria monocytogenes* (**b**).

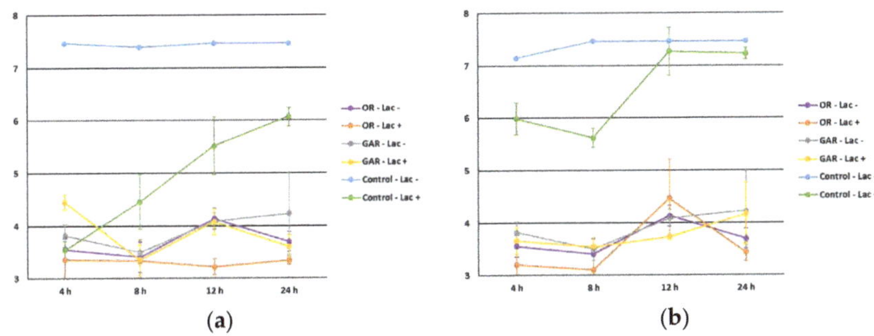

Figure 6. Influence of sodium lactate (absence, Lac−; presence, Lac+) level on the antimicrobial effect of oregano (OR) and garlic (GAR) essential oils against *Salmonella* spp. (**a**) and *Listeria monocytogenes* (**b**).

Regarding sodium lactate, lower microbial counts were observed in the presence of EOs. However, the influence of other factors, namely pH (as previously indicated), could be responsible for the antimicrobial effect and not achievable with respect to the additive.

Regarding commercial phosphates (Figure 7), the antimicrobial effect against *Salmonella* spp., *L. monocytogenes* and *Escherichia coli* in combination with heat treatment was reported by Dickson et al. [36]. Contrary to what was observed for sodium nitrite and sodium lactate, microbiological counts increased along the study period indicating a potential interaction between food phosphates and EOs, decreasing their availability to act against the foodborne pathogens tested. In addition, the potential role of food phosphates as fat and protein emulsifiers in the manufacturing of meat products could decrease the contact of EOs and microorganisms [1]. Thus, it might be hypothesized that a possible interaction between phosphates and the Mueller–Hinton composition (beef extract and casein) decreases the antimicrobial effect of the EOs by the phenomena previously described in the protein interaction.

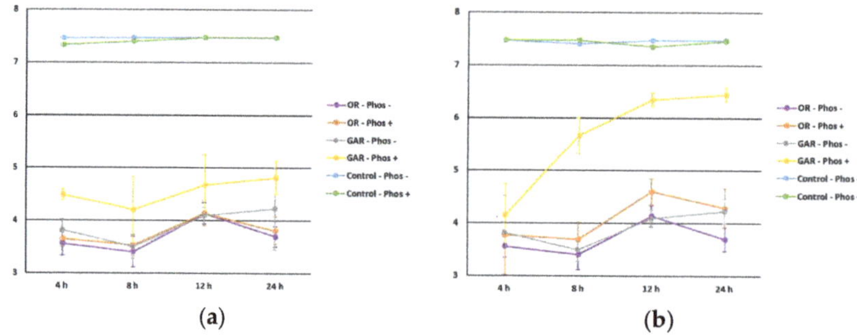

Figure 7. Influence of food phosphates' (absence, Phos−; presence, Phos+) level on the antimicrobial effect of oregano (OR) and garlic (GAR) essential oils against *Salmonella* spp. (**a**) and *Listeria monocytogenes* (**b**).

4. Conclusions

The present work studied the influence of fat, protein, pH, a_w and food additives (sodium nitrite, sodium lactate and food phosphates), individually evaluated, on the antimicrobial effect of oregano and garlic EOs against *Salmonella* spp. and *L. monocytogenes*. Although both EOs presented an important antibacterial effect against the foodborne pathogens tested, the results showed that the presence of fat acts as a barrier to its antimicrobial effect probably due to its dilution on the lipid content of the food model media. The level of protein, pH and a_w did not influence the antimicrobial effect of EOs, although lower microbial counts were observed at the lowest protein level, a_w and pH, respectively. Furthermore, the addition of sodium nitrite and sodium lactate did not influence the inhibitory effect of EOS.

The use of food phosphates could decrease the antimicrobial effect of EOs due to their emulsification properties. The in vitro study of the influence of food characteristic on the antimicrobial effect of EOs could be interesting to the food industry to address the behavior of EOs against specific foodborne pathogens prior their application in foodstuff. Since there is no protocol for evaluating the effect of food additives and essential oils in vitro, the present methodology could be used as an approach for food operators before the utilization of EOs in foodstuffs.

Acknowledgments: This study was supported by the Foundation for Science and Technology of Portuguese Ministry of Education and Science (FCT: Fundacao para a Ciencia e Tecnologia do Ministerio Portugues para a Educacao e Ciencia) through the cofounded project "Portuguese traditional meat products: strategies to improve safety and quality" (PTDC/AGR-ALI/119075/2010) and by a research grant, SFRH/BD/85118/2012, given to the author J. García-Díez.

Author Contributions: Juan García-Díez, Virgílio Falco, Maria João Fraqueza and Luís Patarata conceived of and designed the experiments. Joana Alheiro, Ana Luísa Pinto, Luciana Soares and Virgílio Falco performed the experiments. Juan García-Díez wrote the paper. Virgílio Falco, Maria João Fraqueza and Luís Patarata proofread the text.

Conflicts of Interest: The authors declare no conflict of interest.

References

1. Hyldgaard, M.; Mygind, T.; Meyer, R.L. Essential oils in food preservation: Mode of action, synergies, and interactions with food matrix components. *Front. Microbiol.* **2012**, *3*, 1–24. [CrossRef] [PubMed]
2. Tiwari, B.K.; Valdramidis, V.P.; O'Donnell, C.P.; Muthukumarappan, K.; Bourke, P.; Cullen, P.J. Application of natural antimicrobials for food preservation. *J. Agric. Food Chem.* **2009**, *57*, 5987–6000. [CrossRef] [PubMed]
3. García-Díez, J.; Patarata, L. Behaviour of *Salmonella* spp., *Listeria monocytogenes*, and *Staphylococcus aureus* in Chouriço de Vinho, a dry fermented sausage made from wine-marinated meat. *J. Food Prot.* **2013**, *76*, 588–594. [CrossRef] [PubMed]

4.   Linares, M.B.; Garrido, M.D.; Martins, C.; Patarata, L. Efficacies of Garlic and *L. Sakei* in wine-based marinades for controlling *Listeria monocytogenes* and *Salmonella* spp. inchouriço de vinho, a dry sausage made from wine-marinated pork. *J. Food Sci.* **2013**, *78*, M719–M724. [CrossRef] [PubMed]

5.   Perricone, M.; Arace, E.; Corbo, M.R.; Sinigaglia, M.; Bevilacqua, A. Bioactivity of essential oils: A review on their interaction with food components. *Front. Microbiol.* **2015**, *6*, 76. [CrossRef] [PubMed]

6.   Gutierrez, J.; Barry-Ryan, C.; Bourke, P. Antimicrobial activity of plant essential oils using food model media: Efficacy, synergistic potential and interactions with food components. *Food Microbiol.* **2009**, *26*, 142–150. [CrossRef] [PubMed]

7.   Hierro, E.; Fernandez, M.; de la Hoz, L. Mediterranean Products. In *Handbook of Fermented Meat and Poultry*; Toldrá, F., Hui, Y.H., Astiasaran, I., Sebranek, J., Talon, R., Eds.; Willey Blackwell: West Sussex, UK, 2015; pp. 301–312.

8.   García-Díez, J.; Alheiro, J.; Falco, V.; Fraqueza, M.J.; Patarata, L. Chemical characterization and antimicrobial properties of herbs and spices essential oils against pathogens and spoilage bacteria associated to dry-cured meat products. *J. Essent. Oil Res.* **2016**, *29*, 117–125. [CrossRef]

9.   Talon, R.; Lebert, I.; Lebert, A.; Leroy, S.; Garriga, M.; Aymerich, T.; Drosinos, E.H.; Zanardi, E.; Ianieri, A.; Fraqueza, M.J.; et al. Traditional dry fermented sausages produced in small-scale processing units in Mediterranean countries and Slovakia. 1. Microbial ecosystems of processing environments. *Meat Sci.* **2007**, *77*, 570–579. [CrossRef] [PubMed]

10.   Zaika, L.L. Spices and herbs: Their antimicrobial activity and its determination. *J. Food Saf.* **1987**, *9*, 97–118. [CrossRef]

11.   Clinical and Laboratory Standards Institute. Methods for Dilution Antimicrobial Susceptibility Tests for Bacteria That Grow Aerobically. Approved Standard—Eighth Edition. *Clin. Lab. Stand. Inst.* **2009**, *29*, 2.

12.   Troller, J.A.; Stinson, J.V. Moisture requirements for growth and metabolite production by lactic acid bacteria. *Appl. Environ. Microbiol.* **1981**, *42*, 682–687. [PubMed]

13.   Fratianni, F.; De Martino, L.; Melone, A.; De Feo, V.; Coppola, R.; Nazzaro, F. Preservation of chicken breast meat treated with thyme and balm essential oils. *J. Food Sci.* **2010**, *75*, M528–M535. [CrossRef] [PubMed]

14.   Karabagias, I.; Badeka, A.; Kontominas, M.G. Shelf life extension of lamb meat using thyme or oregano essential oils and modified atmosphere packaging. *Meat Sci.* **2011**, *88*, 109–116. [CrossRef] [PubMed]

15.   Soultos, N.; Tzikas, Z.; Christaki, E.; Papageorgiou, K.; Steris, V. The effect of dietary oregano essential oil on microbial growth of rabbit carcasses during refrigerated storage. *Meat Sci.* **2009**, *81*, 474–478. [CrossRef] [PubMed]

16.   Dussault, D.; Dang Vu, K.; Lacroix, M. In vitro evaluation of antimicrobial activities of various commercial essential oils, oleoresin and pure compounds against food pathogens and application in ham. *Meat Sci.* **2014**, *96*, 514–520. [CrossRef] [PubMed]

17.   Viuda-Martos, M.; Ruiz-Navajas, Y.; Fernández-López, J.; Pérez-Álvarez, J.A. Effect of added citrus fibreand spice essential oils on quality characteristics and shelf-life of mortadella. *Meat Sci.* **2010**, *85*, 568–576. [CrossRef] [PubMed]

18.   Benkeblia, N. Antimicrobial activity of essential oil extracts of various onions (*Allium cepa*) and garlic (*Allium sativum*). *LWT Food Sci. Technol.* **2004**, *37*, 263–268. [CrossRef]

19.   Smith-Palmer, A.; Stewart, J.; Fyfe, L. The potential application of plant essential oils as natural food preservatives in soft cheese. *Food Microbiol.* **2001**, *18*, 463–470. [CrossRef]

20.   Solórzano-Santos, F.; Miranda-Novales, M.G. Essential oils from aromatic herbs as antimicrobial agents. *Curr. Opin. Biotechnol.* **2012**, *23*, 136–141. [CrossRef] [PubMed]

21.   Gil, A.O.; Delaquis, P.; Russo, P.; Holley, R.A. Evaluation of antilisterial action of cilantro oil on vacuum packed ham. *Int. J. Food Microbiol.* **2002**, *73*, 83–92. [CrossRef]

22.   Cava, R.; Nowak, E.; Taboada, A.; Marin-Iniesta, F. Antimicrobial activity of clove and cinnamon essential oils against *Listeria monocytogenes* in pasteurized milk. *J. Food Protect.* **2007**, *70*, 2757–2763. [CrossRef]

23.   Singh, A.; Singh, R.K.; Bhunia, A.K.; Singh, N. Efficacy of plant essential oils as antimicrobial agents against *Listeria monocytogenes* in hotdogs. *LWT Food Sci. Technol.* **2003**, *36*, 787–794. [CrossRef]

24.   Drumm, T.D.; Spanier, A.M. Changes in the content of lipid autoxidation and sulfur-containing compounds in cooked beef during storage. *J. Agric. Food Chem.* **1991**, *39*, 336–343. [CrossRef]

25. Abdollahzadeh, E.; Rezaei, M.; Hosseini, H. Antibacterial activity of plant essential oils and extracts: The role of thyme essential oil, nisin, and their combination to control *Listeria monocytogenes* inoculated in minced fish meat. *Food Control* **2014**, *35*, 177–183. [CrossRef]

26. Baranauskien, R.; Venskutonis, P.R.; Dewettinck, K.; Verhe, R. Properties of oregano (*Origanum vulgare* L.), citronella (*Cymbopogonnardus* G.) and marjoram (*Majorana hortensis* L.) flavors encapsulated into milk protein-based matrices. *Food Res. Int.* **2006**, *39*, 413–425. [CrossRef]

27. Skandamis, P.; Tsigarida, E.; Nychas, G.J. Ecophysiological attributes of *Salmonella typhimurium* in liquid culture and within a gelatin gel with or without the addition of oregano essential oil. *World J. Microbiol. Biotechnol.* **2000**, *16*, 31–35. [CrossRef]

28. Burt, S. Essential oils: Their antibacterial properties and potential applications in foods—A review. *Int. J. Food Microbiol.* **2004**, *94*, 223–253. [CrossRef] [PubMed]

29. Mølbak, K.; Olsen, J.E.; Wegener, H.C. Salmonella infections. In *Foodborne Infections and Intoxications*; Cliver, D.O., Potter, M., Riemann, H.P., Eds.; Academic Press: Washington, DC, USA, 2011; pp. 57–136.

30. Pagotto, F.; Corneau, N.; Farber, J. *Listeria monocytogenes*. In *Foodborne Infections and Intoxications*; Cliver, D.O., Potter, M., Riemann, R.P., Eds.; Academic Press: Washington, DC, USA, 2011; pp. 312–340.

31. Gandhi, M.; Chikindas, M.L. Listeria: A foodborne pathogen that knows how to survive. *Int. J. Food Microbiol.* **2007**, *113*, 1–15. [CrossRef] [PubMed]

32. Skandamis, P.N.; Nychas, G.J.E. Development and evaluation of a model predicting the survival of *Escherichia coli* O157: H7 NCTC 12900 in homemade eggplant salad at various temperatures, pHs, and oregano essential oil concentrations. *Appl. Environ. Microbiol.* **2000**, *66*, 1646–1653. [CrossRef] [PubMed]

33. Negi, P.S. Plant extracts for the control of bacterial growth: Efficacy, stability and safety issues for food application. *Int. J. Food Microbiol.* **2012**, *156*, 7–17. [CrossRef] [PubMed]

34. Cassens, R.G. Composition and safety of cured meats in the USA. *Food Chem.* **1997**, *59*, 561–566. [CrossRef]

35. Gonzalez, B.; Díez, V. The effect of nitrite and starter culture on microbiological quality of "chorizo"—A Spanish dry cured sausage. *Meat Sci.* **2002**, *60*, 295–298. [CrossRef]

36. Dickson, J.S.; Cutter, C.G.; Siragusa, G.R. Antimicrobial effects of trisodium phosphate against bacteria attached to beef tissue. *J. Food Prot.* **1994**, *57*, 952–955. [CrossRef]

 *foods*

Article

Rapid Prediction of Moisture Content in Intact Green Coffee Beans Using Near Infrared Spectroscopy

Adnan Adnan [1], Dieter von Hörsten [2], Elke Pawelzik [1] and Daniel Mörlein [3,*

[1]   Division Quality of Plant Products, Department of Crop Sciences, University of Goettingen, Carl-Sprengel-Weg 1, 37075 Goettingen, Germany; adnan.adnan@stud.uni-goettingen.de (A.A.); epawelz@gwdg.de (E.P.)

[2]   Institute for Application Techniques in Plant Protection, Julius Kühn Institute, Messeweg 11/12, 38140 Braunschweig, Germany; dieter.von-hoersten@jki.bund.de

[3]   Department of Animal Sciences, University of Goettingen, Albrecht-Thaer-Weg 3, D-37075 Goettingen, Germany

*   Correspondence: daniel.moerlein@agr.uni-goettingen.de; Tel.: +49-551-39-5611; Fax: +49-551-39-5587

Academic Editors: Theo H. Varzakas and Charalampos Proestos
Received: 10 March 2017; Accepted: 17 May 2017; Published: 19 May 2017

Abstract: Moisture content (MC) is one of the most important quality parameters of green coffee beans. Therefore, its fast and reliable measurement is necessary. This study evaluated the feasibility of near infrared (NIR) spectroscopy and chemometrics for rapid and non-destructive prediction of MC in intact green coffee beans of both *Coffea arabica* (Arabica) and *Coffea canephora* (Robusta) species. Diffuse reflectance (log 1/R) spectra of intact beans were acquired using a bench top Fourier transform NIR instrument. MC was determined gravimetrically according to The International Organization for Standardization (ISO) 6673. Samples were split into subsets for calibration (n = 64) and independent validation (n = 44). A three-component partial least squares regression (PLSR) model using raw NIR spectra yielded a root mean square error of prediction (RMSEP) of 0.80% MC; a four component PLSR model using scatter corrected spectra yielded a RMSEP of 0.57% MC. A simplified PLS model using seven selected wavelengths (1155, 1212, 1340, 1409, 1724, 1908, and 2249 nm) yielded a similar accuracy (RMSEP: 0.77% MC) which opens the possibility of creating cheaper NIR instruments. In conclusion, NIR diffuse reflectance spectroscopy appears to be suitable for rapid and reliable MC prediction in intact green coffee; no separate model for Arabica and Robusta species is needed.

Keywords: quality; rapid methods; infrared spectroscopy; *Coffea arabica* (Arabica); *Coffea canephora* (Robusta); chemometrics

---

1. Introduction

Moisture content (MC) is one of the most important quality parameters of green coffee beans. Most importing and exporting countries regulate MC as one of the quality standards for green coffee beans. The safety range for MC is 8.0–12.5%, based on fresh matter [1–3]. MC outside the safety range impairs the bean quality and safety. Beans with a MC above 12.5% are not allowed to be shipped and traded [4]. MC below 8% causes shrunken beans and an unwanted appearance [5], whereas MC above 12.5% facilitates fungal growth and mycotoxin production (e.g., ochratoxin A) that are risks to human health [6,7].

Coffee is harvested in the form of ripe berries and has a MC of more than 60% [8]. These ripe berries are processed through several steps of (wet or dry) postharvest treatments resulting in green coffee beans. Farmers generally dry the beans under the sun. The dried beans often do not meet the standard requirements for MC, resulting in a lower price [9]. For example, green beans obtained in the Bengkulu Province of Indonesia had a MC of 10.1–18.6% [10] and those in West Nusa Tenggara Province had a MC of 11.0–14.1% [11].

MC control is also important for the storability of the beans. An inappropriate storage environment (e.g., non-aerated silos and bag storage) affects MC fluctuation. The MC of green coffee beans stored in non-aerated silos increased up to 15.4% during rainy season. This moisture increase leads to the accumulation of glucose and an unpleasant taste in the beverage [12].

Furthermore, MC is crucial before the roasting process. The same roasting temperature and time with different MCs can result in different quality attributes—like color, density, and aroma—of the end product [13]. Consequently, an identical MC of green coffee beans is important for the roasting procedure in order to produce a consistent quality of roasted beans.

Therefore, a fast and accurate determination of MC in green coffee beans is vital. Up to date, the standard method for determining MC is the gravimetric method, where a drying chamber with a certain temperature and time is used to dry the beans and afterwards the mass loss is calculated. International standards for MC measurement of green coffee beans are The International Organization for Standardization (ISO) 1446, 1447, and 6673 [3,14]. Thereof, ISO 6673 which requires less preparation and the shortest drying time (105 °C for 16 h) is widely accepted as a reference method among importing and exporting countries. Apparently these gravimetric methods do not suffice when the information on MC is needed instantly [5] which is why we researched alternative methods.

Near infrared spectroscopy (NIRS) has been widely investigated for rapid, often non-destructive, determination of the compositional and quality traits of agricultural products. For example, previous work in our group predicted rapid and non-destructive analysis of mango quality attributes using NIRS and chemometrics [15]. NIRS makes use of the fact that near infrared (NIR) radiation in the range of 780–2500 nm predominantly interacts with hydrogen bonds—e.g., O–H, C–H, N–H, S–H. NIR radiation that hits a sample may be transmitted, absorbed, or reflected, this depends on the chemical composition and physical factors of the sample. The intensity of transmitted, absorbed, or reflected radiation is then recorded by NIRS [16,17].

Specific wavelengths (1450 and 1940 nm) were identified to be highly correlated with water content [3,18,19]. Predicting MC using NIRS in any agricultural product is more complex and should not be based on wavelengths limited to 1450 and 1940 nm. MC does not only reflect water, but also loss of volatile compounds during drying [3]. In fact, NIR has some disadvantages, e.g., overlapping of wavelengths that correspond to specific organic compounds, and scattering problems [16,20]. It is therefore necessary to carefully develop calibration models for NIR based predictions [18,19].

Prediction of MC by NIRS has been developed over years for many agricultural products [21]. A regression model was developed to predict MC in (ground) green coffee bean (*Coffea arabica* from Brazil) based on NIR diffuse reflectance (log 1/R) spectra [22]. To the best of our knowledge, this is the first study investigating the prediction of moisture content in intact green coffee beans of both *Coffea arabica* (Arabica) and *Coffea canephora* (Robusta) species by near infrared spectroscopy (NIRS) and chemometrics. The main goal of this paper was to study the feasibility of near infrared spectroscopy (NIRS) to predict moisture content (MC) in intact green coffee beans. We developed and validated calibration models based on diffuse reflectance spectra which were obtained using a benchtop near infrared instrument. Our decision to involve both Arabica and Robusta species stems from the fact that both species are commercially important but vary in their chemical composition. Furthermore, we used intact green beans such as no sample preparation would be needed—yet such an approach has not been documented. The results are especially relevant for those involved in coffee trading, production, and quality control. We also demonstrate the possibility of creating a simple NIR instrument which only uses a few important wavelengths to predict MC, rather than employing the full NIR spectrum.

2. Materials and Methods

A schematic representation of the experimental set up is given in the supplementary information (Figure S1).

2.1. Materials

Green Arabica and Robusta coffee beans that were harvested in 2013 were obtained from a local trading company in Indonesia. The materials were divided into separate sample sets for calibration and validation purposes (Table 1). The beans were placed in an open plastic box with the size of 15.5 × 11 × 6 cm, and were stored in a climatic chamber (Rumed® type 1301, Rubarth Apparate GmbH, Laatzen, Germany) at 25 °C and a relative humidity range of 30–85%, in order to obtain a broad range of MC within 6–22%. Upon equilibration, samples were removed from the climatic chamber to record diffuse reflectance (log 1/R) data. Immediately thereafter, MC of the beans was determined.

Table 1. Characteristics of the coffee samples including species and origin.

No.	Purpose	Species	Origin
1			West Nusa Tenggara
2		Arabica	South Sulawesi
3			Aceh
4	Calibration		South Sumatera
5		Robusta	Bali
6			East Java
7			North Sumatera
8		Arabica	West Java
9			North Sumatera
10	Validation		South Sumatera
11		Robusta	East Java
12			Bengkulu

2.2. Near-Infrared Spectroscopy

A bench top Fourier transform near infrared (FT-NIR) instrument with sample cup rotation (Thermo Nicolet Antaris MDS, Thermo Fisher, Waltham, MA, USA) was used to acquire diffuse reflectance spectra (log 1/R) of bulk samples of green coffee beans (40 g) on a Petri dish with a diameter of 7 cm.

Spectra were collected according to a workflow developed using the software Result Integration Software (Result™ version 3.0, Thermo Fisher, Waltham, MA, USA). Internal background spectra were collected once every hour. High resolution diffuse reflectance (log 1/R) spectra at a wavelength range of 1000 to 2500 nm with 2 nm intervals were recorded as the averages of 64 scans. Thus, the spectra consisted of 1557 data points. Three replicates were acquired per sample and the spectra were averaged before further calculations. In total, this resulted in 108 spectra of 12 samples differing in moisture content, species, and origin.

2.3. Moisture Content Determination

MC (% wet basis) was determined was based on ISO 6673 [3]. A forced air electrical oven (Thermicon P® type UT6120, Heraeus Instruments GmbH, Hanau, Germany) was used to dry approximately 10 g whole green coffee beans in open glass petri dishes (diameter: 14 cm, height: 2.3 cm) at 105 ± 1 °C for 16 h. Samples were limited to six origins with two replications per drying cycle in order to maintain an equal amount of mass loss during drying. The petri dishes were closed with glass lids immediately after drying had completed, and then they were stored in desiccators for 1 h in order to cool down the samples to ambient temperature. Finally, samples were weighted (Type LP 620 S, Sartorius AG, Göttingen, Germany) to calculate MC based on weight loss; data are given as the average from two replications (Equation (1)). Across all samples, average standard deviation of replicate MC determinations was 0.21% MC (Median: 0.08% MC).

$$MC = \frac{W_w - W_d}{W_w} \tag{1}$$

where MC is the moisture content (%) of green coffee beans (wet basis), W_W is the wet weight of the sample, and W_d is the weight of the sample after drying.

2.4. Data Processing

The statistical software (The Unscrambler® X version 10.2 Network Client, CAMO software AS, Oslo, Norway) was used for further processing of the spectral data. Regression models to predict MC in green coffee beans were developed with a subset of calibration samples ($n = 64$), and then the models were tested using the subset of validation samples ($n = 44$) to evaluate the accuracy.

Firstly, spectral outliers were identified using Principal Component Analysis (PCA) and Hotelling's T^2 ellipse 5% plot, based on all samples' ($n = 108$) raw spectra. Afterwards, several pre-processing methods were applied to compensate the disadvantages of NIR, e.g., the scattering and material size [16,23]. In detail, smoothing (moving average, Gaussian filter, median filter) window size of 3, 7, 11, 15, 19; Savitsky–Golay derivative (First derivative, two polynomial order; second derivative, two polynomial order; third derivative, three polynomial order) window size of 3, 7, 11, 15, 19; normalization (area, mean); baseline correction (baseline offset, linear baseline correction); standard normal variate (SNV); orthogonal signal correction (OSC) (non-linear iterative partial least squares algorithm, number of component 1); multiplicative scatter correction (MSC) (full MSC model); and extended multiplicative scatter correction (EMSC) were applied. Subsequently, the models were compared in terms of prediction accuracy and model robustness (number of latent variables). MSC and EMSC were applied to the calibration data. Upon model validation, the processing was also applied to the validation data set prior to prediction.

Calibration models were developed using both partial least squares regression (PLSR) and multiple linear regression (MLR). For PLSR, the full spectra (1557 wave numbers, mean centered) were used. Full cross validation was applied to estimate calibration errors. Regression coefficients were obtained from PLSR to determine the important wavelengths, i.e., those that correlated most to MC. A subset of selected wavelengths was then used as an input for full rank MLR and PLS regression to identify the most parsimonious yet robust model. Leverage correction was applied with MLR to estimate calibration errors. The calibration models derived from PLSR and MLR were evaluated by the number of latent variables (LVs), R^2 of calibration, R^2 of cross validation, root mean square error of calibration (RMSEC), and root mean square error of cross validation (RMSECV). Finally, all models were validated in terms of their prediction accuracy using a separate validation data set. Parameters used were R^2 of prediction, root mean square error of prediction (RMSEP), standard error of prediction (SEP), bias, and residual predictive deviation (RPD) [22,24].

3. Results

3.1. Spectral Properties, Outliers, and Effect of Pre-Processing

According to an initial PCA using all raw spectra and projection of the Hotelling's T^2 ellipse, four samples were suspected as spectral outliers (Figure 1). Subsequent modeling with and without these potential outliers, respectively revealed that model accuracy was not significantly affected. Thus, the suspected outliers were not excluded.

Inspection of the raw data also revealed that the NIR diffuse reflectance spectra of intact green coffee beans are influenced by scatter (Figure 2a). Several pre-processing methods were applied to eliminate the scatter. Application of EMSC proved to improve the prediction accuracy; the EMSC corrected spectra are shown in Figure 2b. Inspection of EMSC corrected spectra indicated that several wavelength regions reflect the chemical information regarding moisture content.

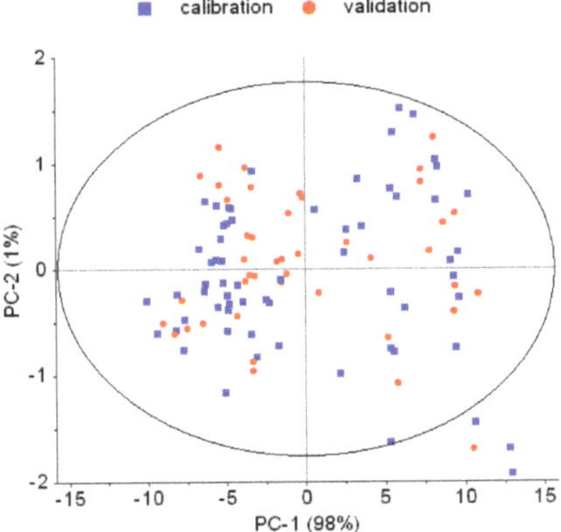

Figure 1. Score plot of principal component analysis (PCA) using raw infrared spectra (log 1/R) with Hotelling's T^2 ellipse for outlier inspection. Calibration samples (squares) and validation samples (circles) are marked accordingly. PC: principal component.

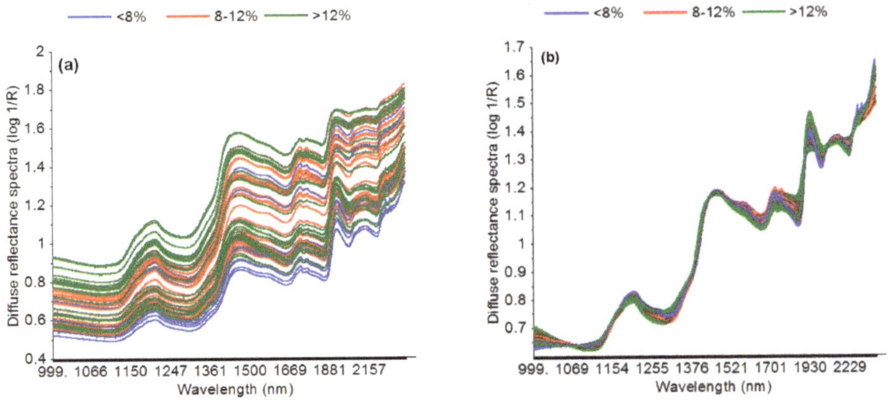

Figure 2. Diffuse reflectance spectra (log 1/R) of calibration model. Raw spectra (**a**); EMSC (extended multiplicative scatter) corrected spectra (**b**).

3.2. Prediction of Moisture Content from NIR Reflectance Spectra

Several preprocessing methods were applied to build the model (see Section 2.4). Nevertheless, none of the preprocessing methods yielded a better accuracy than models using raw data. Selected results of the various chemometric approaches to predict MC from NIR reflectance spectra are given in Table 2. The most parsimonious PLSR model on the full spectral range was achieved using raw spectra and three latent variables. Its prediction accuracy was, however, somewhat compromised when using the independent validation data set. Using the EMSC corrected spectra instead of the raw spectra yielded a similar R^2 while the prediction errors were comparably low both for the calibration and the validation data set. Yet, this model used four latent variables, e.g., it was less parsimonious compared to the model based on raw data.

Table 2. Statistical parameters of the developed prediction models for moisture content (MC) in green coffee beans using near infrared spectra.

Model	Parameter	Full Spectral Range PLSR		Spectral Subset	
		Raw	EMSC	Raw (MLR)	Raw (PLS)
	LVs	3	4	n/a	3
	R^2 calibration	0.9834	0.9850	0.9839	0.9743
Calibration	R^2 cross validation	0.9802	0.9811	0.9779	0.9698
	RMSEC (% MC)	0.52	0.49	0.51	0.65
	RMSECV (% MC)	0.58	0.56	0.60	0.71
	R^2 prediction	0.9641	0.9817	0.9632	0.9669
Prediction	RMSEP (% MC)	0.80	0.57	0.93	0.77
	Bias (% MC)	0.42	0.28	0.45	0.39
	RPD	6.21	8.53	3.47	6.39

PLSR: partial least squares regression using full spectral range (1000 to 2500 nm, 1557 data points); MLR/PLS: multiple linear and partial least squares regression using selected wavenumbers (1155, 1212, 1340, 1409, 1724, 1908, and 2249 nm); LVs: Latent variables (for PLS only); R^2: the coefficient of determination; RMSEC: root mean square error of valibration; RMSECV: root mean square error of cross validation; RMSEP: root mean square error of prediction; SEP: standard error of prediction; RPD: residual predictive deviation; n/a: not applicable; MC: moisture content.

Principal components (PCs) 1 and 2 of the PLSR model based on raw spectra explain 99% of spectral data variance and 51% of MC variance; a clear separation of Arabica and Robusta species is to be seen (Figure 3a). PC 2 and 3 together explain 94% of MC variance (Figure 3b).

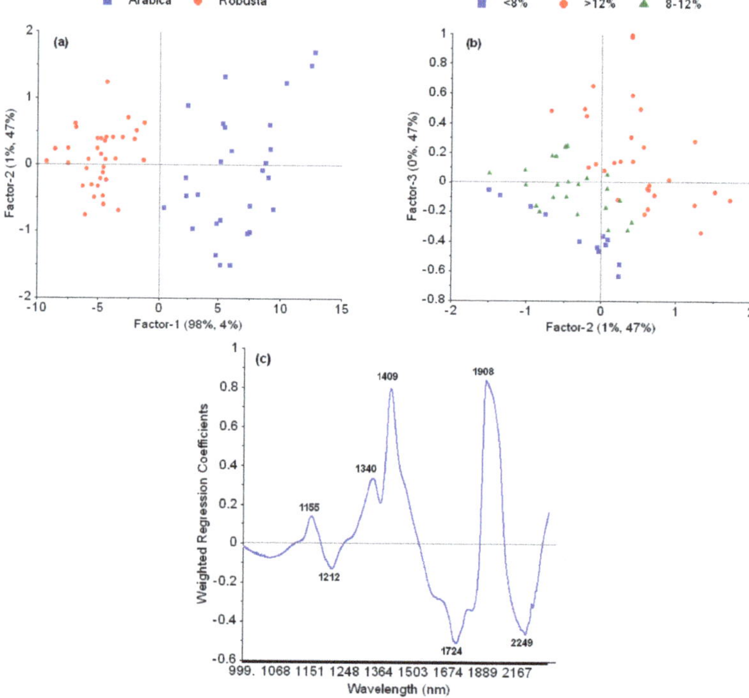

Figure 3. Score plots of PLSR for moisture content prediction based on raw diffuse reflectance (log 1/R) near infrared spectra. A distinct clustering of Arabica and Robusta coffee samples is observed when displaying PC 1 vs. PC2 (herein: factor-1 and factor-2) (**a**); Sample allocation is following moisture content indicating the importance of PC 2 and 3 for moisture prediction (**b**); Weighted regression coefficients obtained from PLSR using raw spectra (**c**).

Weighted regression coefficients obtained from PLSR on raw data (Figure 3c) were then used to study whether the model could be even simplified. Note that weighted and raw regression coefficients are the same as long as spectral data are not scaled but only mean centered; this was applied here. Seven wavelengths were selected due to their regression weights. That is, the intensities of 1155, 1212, 1340, 1409, 1724, 1908, and 2249 nm were used as input data to develop a MLR calibration model. Thus, a similarly accurate model was obtained (Table 2); the prediction error for the validation test set was significantly lower ($p < 0.05$) for the MLR model (0.93% MC) as compared to the EMSC model using raw data (0.57% MC). The resulting MLR model is given in Equation (2).

$$\text{MC (\%)} = -4.20 + 115.02\,(V1) + 0.40\,(V2) - 116.18\,(V3) \\ + 76.16\,(V4) - 97.72\,(V5) + 63.76\,(V6) - 17.59\,(V7) \tag{2}$$

where, V1 to V7 are the intensities of the wavelengths 1155, 1212, 1340, 1409, 1724, 1908, and 2249 nm, respectively. When subjecting this spectral subset to PLS, the predictive ability of a three LV model was even improved as compared to the full-rank MLR model (Table 2); its prediction error (0.77% MC) was significantly lower than the MLR model ($p = 0.015$). It is, however, not significantly different from the PLSR model using raw data ($p > 0.05$).

PLSR and MLR using raw spectral data yielded a good correlation of reference versus predicted MC (Figure 4a,b). Also, the model's bias is close to the error of the reference method (0.21% MC, see 2.3.).

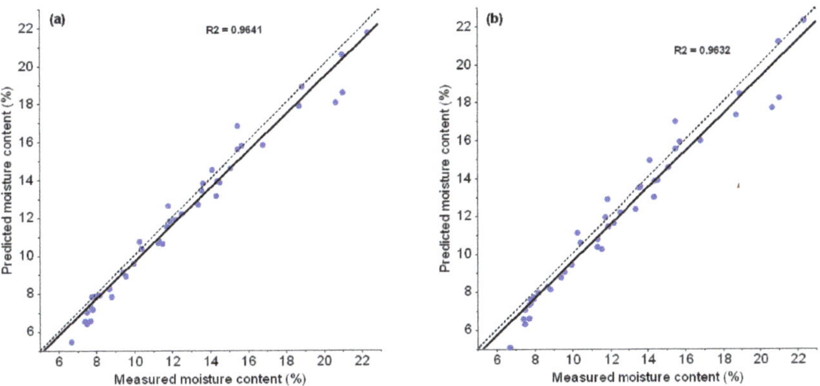

Figure 4. Predicted vs. measured moisture content of green coffee beans based on raw diffuse reflectance (log 1/R) near infrared spectra. (**a**) PLSR; (**b**) MLR.

4. Discussion

4.1. Outliers and Effect of Pre-Processing

For outlier detection, PCA and subjection of the Hotelling's T^2 ellipse along with residuals and influence plot, and Q-residuals plot, were used which are common approaches in multivariate analysis. Identifying true outliers is important to prevent false inferences [25]. In this experiment, four samples were suspected to be outliers (Figure 1). Explained spectral variance (PC1 + PC2) based on diffuse raw data reflectance (log 1/R) was 99%. Elimination of suspected outliers did not increase the explained variance. Further comparisons of PLSR with and without the suspected outliers yielded only very slight improvement in R^2 which indicates that the suspected outliers were no real outliers. Similarly, Morales-Medina and Guzmán [26] examined multivariate data using Hotelling's T^2 ellipse. They also decided to not exclude the suspected outliers because they did not significantly affect the explained variance found through PCA.

Various pre-processing methods were applied to the raw spectra. This aims at reducing noise and improving the accuracy of the prediction model [27]. EMSC was effective to remove scatter which was shown also in other studies [28]. Accordingly, the prediction errors were the lowest when using EMSC corrected data for PLSR (Table 2). The resulting model, however, was surprisingly less parsimonious, i.e., it needed one more latent variable. Pizarro et al. [27] also reported that none of the pre-processing methods studied (first and second derivation, MSC, standard normal variate) improved the prediction for ash and lipid content in roasted coffee significantly as compared to using raw data; only OSC and direct orthogonal signal correction (DOSC) enhanced the model performance remarkably.

4.2. Prediction of Moisture Content Using NIR Infrared Spectra

Raw spectra were selected as an input to build the final PLSR model because this resulted in the lowest number of latent variables, the highest R^2 and lowest root mean square error compared to other pre-processing methods (Table 2). A model with these criteria is preferable. Kamruzzaman et al. [29] also considered the number of latent variables together with R^2 and prediction errors to select the most appropriate model for prediction of water, fat, and protein content in lamb meat. Both the robustness and the predictive ability of a given model are of importance. If one considers only R^2, RMSEP, or RPD, which reflect the predictive ability, likely models using more latent variables would be preferred over models using less latent variables. In terms of robustness, however, a model using less latent variables is less prone to overfitting than a model using more latent variables.

Further examination of the PLSR score plots (based on raw spectra) revealed a distinct clustering of Arabica and Robusta samples on the first latent variable, explaining 98% in the spectral data variance but only 4% of moisture variance (Figure 3a). To understand this clustering, the loading weights of the first LV were inspected. As a result, important wavelengths are related to several chemical compounds, e.g., caffeine, chlorogenic acid, lipids, protein and amino acids, sucrose, carbohydrates, trigonelline and, of course, water [30]. These compounds were shown to vary between species which is why their spectral contributions can be used to discriminate between species [31,32] Using PC 2 and 3 which together explain 94% of moisture variance, samples are allocated according to moisture content levels (Figure 3b). Thus, a three component PLSR model allows prediction of moisture content on both Arabica and Robusta species. The advantages of inputting raw spectra rather than pre-processed spectra firstly reduces the complexity of calculations and therefore secondly reduces the computation time. These advantages will be useful for online and real time prediction in the future.

The statistical parameters of calibration and prediction accuracy were similar for the developed PLSR models, especially for the model based on EMSC corrected spectra. This indicates that the PLSR model is robust in terms of predicting unknown samples accurately. We also investigated PLSR models based on raw spectra within individual species. However, the results were not better than the PLSR model which was developed across species. The PLSR model obtained in this experiment resulted in a similar accuracy compared to what was reported by Morgano et al. [22]. That study predicted the MC of green Arabica coffee beans, based on smoothed spectra, which yielded an R^2 of calibration = 0.507, R^2 of validation = 0.669, and RMSEV of 0.55% MC (R^2 recalculated from r).

Even simplified MLR and PLS models were built using selected wavelengths based on their relative importance in the PLSR model. This experiment showed that near infrared diffuse reflectance intensities at 1155, 1212, 1340, 1409, 1724, 1908, and 2249 nm highly correspond to MC (Figure 3c). According to Ribeiro et al. [30], these wavelengths are related to the absorbance of the second overtone of C–H, first combination overtone of C–H, first overtone of O–H and N–H, second overtone of C=O, and combination of O–H and N–H, respectively. Obviously, these wavelengths are not exactly located at the water bands which indicate that it may well be useful to apply indirect relationships in prediction models. Plus, it was shown that the degradation of organic components during drying for MC determination needs to be considered. Reh et al. [3] proved that, using ISO 6673, the beans lose 0.39% of their mass besides water. Thus, MC is calculated as a sum of extracted water and mass losses of other compounds. Similarly, Pan et al. [33] found that MC in beet slices highly corresponded to spectral intensities at 968, 1078, and 1272 nm, i.e., not exactly located at the water bands.

The MLR model, as well as the PLS model based on the spectral subset, yielded a good accuracy both for calibration and validation thus proving their robustness (Figure 4b). The biases measured by PLSR and MLR were close to the method error of determining moisture content based on ISO 6673. Moreover, the ratio of the standard deviation of the target variable and the SEP of a given model, commonly referred to as RPD (residual predictive deviation), is often used to assess the performance of prediction models; higher RPD values indicate a better predictive performance [24]. Here, the models yielded RPD values of about 3 to 8 (Table 2) which is considered good [34]. This shows the potential of near infrared spectroscopy to replace the reference method when a fast and non-destructive prediction is needed, e.g., when trading or for in-line process control.

Finally, the remarkable reduction of variables without a relevant loss of accuracy opens the possibility of creating a simple NIR instrument which only uses a few important wavelengths to predict MC, rather than employing the full NIR spectrum. Specific LED light sources emitting only selected wavelengths can potentially reduce the costs of an NIR instrument.

5. Conclusions

The results indicate that a fast, non-destructive prediction of moisture content in intact green coffee beans is feasible using near infrared diffuse reflectance spectroscopy. EMSC effectively reduces scatter apparent in raw spectra. Thus, the prediction accuracy using EMSC corrected spectra is improved at the cost of a somewhat less parsimonious model. A simplified model based on only seven selected wavelengths points to the possibility of a cheaper instrumentation. The calibration model can be applied for both Arabica and Robusta species. In conclusion, NIR is deemed feasible to replace gravimetric methods for routine applications where a timely result may outweigh the loss of accuracy as compared to the drying methods.

Supplementary Materials: The following are available online at www.mdpi.com/2304-8158/6/38/s1, Figure S1: Schematic representation of the experimental set up.

Acknowledgments: We would like to thank the Ministry of Agriculture of the Republic of Indonesia for granting the scholarship to Adnan.

Author Contributions: Adnan and Dieter von Hörsten conceived and designed the experiments; Adnan performed the experiments; Adnan and Daniel Mörlein analyzed the data; Daniel Mörlein contributed analysis tools; Adnan and Daniel Mörlein wrote the paper. Elke Pawelzik and Daniel Mörlein supervised the study. All authors read and approved the manuscript.

Conflicts of Interest: The authors declare no conflict of interest. The founding sponsors had no role in the design of the study; in the collection, analyses, or interpretation of data; in the writing of the manuscript, or in the decision to publish the results.

References

1. International Coffee Organization (ICO). National quality standards. In Proceedings of the Promotion and Market Development Committee 6th Meeting, Belo Horizonte, Brazil, 9 September 2013. Available online: http://www.ico.org/documents/cy2012-13/pm-29e-quality-standards.pdf (accessed on 1 February 2017).
2. Pittia, P.; Nicoli, M.C.; Sacchetti, G. Effect of moisture and water activity on textural properties of raw and roasted coffee beans. *J. Texture Stud.* **2007**, *38*, 116–134. [CrossRef]
3. Reh, C.; Gerber, A.; Prodolliet, J.; Vuataz, G. Water content determination in green coffee—Method comparison to study specificity and accuracy. *Food Chem.* **2006**, *96*, 423–430. [CrossRef]
4. Van Hilten, H.J.; Fisher, P.J.; Wheeler, M.A. *The Coffee Exporter's Guide*, 3rd ed.; International Trade Centre UNCTAD/GATT: Geneva, Switzerland, 2011.
5. Gautz, L.D.; Smith, V.E.; Bittenbender, H.C. Measuring coffee bean moisture content. *Eng. Noteb.* **2008**, *3*, 1–3.
6. Palacios-Cabrera, H.; Taniwaki, M.H.; Menezes, H.C.; Iamanaka, B.T. The production of ochratoxin A by Aspergillus ochraceus in raw coffee at different equilibrium relative humidity and under alternating temperatures. *Food Control* **2004**, *15*, 531–535. [CrossRef]

7. Pardo, E.; Ramos, A.; Sanchis, V.; Marin, S. Modelling of effects of water activity and temperature on germination and growth of ochratoxigenic isolates of on a green coffee-based medium. *Int. J. Food Microbiol.* **2005**, *98*, 1–9. [CrossRef] [PubMed]

8. Finzer, J.R.D.; Limaverde, J.R.; Freitas, A.O.; Júnior, J.L.; Sfredo, M.A. Drying of coffee berries in a vibrated tray dryer operated with solids recycle and single-stage. *J. Food Process Eng.* **2003**, *26*, 207–222. [CrossRef]

9. Subedi, R.N. Comparative analysis of dry and wet processing of coffee with respect to quality and cost in kavre district, Nepal: A case of panchkhal village. *Int. Res. J. Appl. Basic Sci.* **2011**, *2*, 181–193.

10. Yani, A. Fungal infection at coffee beans during primary processing (case study in Bengkulu Province). *Akta Agrosia* **2008**, *11*, 87–95.

11. Aklimawati, L.; Mawardi, S. Characteristics of Quality Profile and Agribusiness of Robusta Coffee in Tambora Mountainside, Sumbawa. *Pelita Perkeb. Coffee Cocoa Res. J.* **2014**, *30*, 159–180.

12. Bucheli, P.; Meyer, I.; Pittet, A.; Vuataz, G.; Viani, R. Industrial storage of green Robusta coffee under tropical conditions and its impact on raw material quality and ochratoxin A content. *J. Agric. Food Chem.* **1998**, *46*, 4507–4511. [CrossRef]

13. Baggenstoss, J.; Poisson, L.; Kaegi, R.; Perren, R.; Escher, F. Roasting and aroma formation: Effect of initial moisture content and steam treatment. *J. Agric. Food Chem.* **2008**, *56*, 5847–5851. [CrossRef] [PubMed]

14. Mendonça, J.C.F.; Franca, A.S.; Oliveira, L.S. A comparative evaluation of methodologies for water content determination in green coffee. *LWT - Food Sci. Technol.* **2007**, *40*, 1300–1303. [CrossRef]

15. Munawar, A.A.; von Hörsten, D.; Wegener, J.K.; Pawelzik, E.; Mörlein, D. Rapid and non-destructive prediction of mango quality attributes using Fourier transform near infrared spectroscopy and chemometrics. *Eng. Agric. Environ. Food* **2016**, *9*, 208–215. [CrossRef]

16. Blanco, M.; Villarroya, I. NIR spectroscopy: A rapid-response analytical tool. *TrAC Trends Anal. Chem.* **2002**, *21*, 240–250. [CrossRef]

17. Nicolaï, B.M.; Beullens, K.; Bobelyn, E.; Peirs, A.; Saeys, W.; Theron, K.I.; Lammertyn, J. Nondestructive measurement of fruit and vegetable quality by means of NIR spectroscopy: A review. *Postharvest Biol. Technol.* **2007**, *46*, 99–118. [CrossRef]

18. Isengard, H.D. Rapid water determination in foodstuffs. *Trends Food Sci. Technol.* **1995**, *6*, 155–162. [CrossRef]

19. Isengard, H.-D. Water content, one of the most important properties of food. *Food Control* **2001**, *12*, 395–400. [CrossRef]

20. Barbin, D.F.; Felicio, A.L.d.S.M.; Sun, D.-W.; Nixdorf, S.L.; Hirooka, E.Y. Application of infrared spectral techniques on quality and compositional attributes of coffee: An overview. *Food Res. Int.* **2014**, *61*, 23–32. [CrossRef]

21. Büning-Pfaue, H. Analysis of water in food by near infrared spectroscopy. *Food Chem.* **2003**, *82*, 107–115. [CrossRef]

22. Morgano, M.A.; Faria, C.G.; Ferrão, M.F.; Bragagnolo, N.; Ferreira, M.M.d.C. Determination of moisture in raw coffee by near infra-red reflectance spectroscopy and multivariate regression. *Food Sci. Technol. Camp.* **2008**, *28*, 12–17.

23. Esteban-Díez, I.; González-Sáiz, J.M.; Pizarro, C. An evaluation of orthogonal signal correction methods for the characterisation of arabica and robusta coffee varieties by NIRS. *Anal. Chim. Acta* **2004**, *514*, 57–67. [CrossRef]

24. Fearn, T. Assessing Calibrations: SEP, RPD, RER and R 2. *NIR News* **2002**, *13*, 12–13. [CrossRef]

25. Shabbak, A.; Midi, H. An Improvement of the Hotelling T 2 Statistic in Monitoring Multivariate Quality Characteristics. *Math. Probl. Eng.* **2012**, *2012*, 1–15. [CrossRef]

26. Morales-Medina, G.; Guzmán, A. Prediction of density and viscosity of Colombian crude oils from chromatographic data. *CTF-Cienc. Tecnol. Futuro* **2012**, *4*, 57–73.

27. Pizarro, C.; Esteban-Díez, I.; Nistal, A.-J.; González-Sáiz, J.-M. Influence of data pre-processing on the quantitative determination of the ash content and lipids in roasted coffee by near infrared spectroscopy. *Anal. Chim. Acta* **2004**, *509*, 217–227. [CrossRef]

28. Sørensen, K.M.; Petersen, H.; Engelsen, S.B. An On-Line Near-Infrared (NIR) Transmission Method for Determining Depth Profiles of Fatty Acid Composition and Iodine Value in Porcine Adipose Fat Tissue. *Appl. Spectrosc.* **2012**, *66*, 218–226. [CrossRef] [PubMed]

29. Kamruzzaman, M.; ElMasry, G.; Sun, D.-W.; Allen, P. Non-destructive prediction and visualization of chemical composition in lamb meat using NIR hyperspectral imaging and multivariate regression. *Innov. Food Sci. Emerg. Technol.* **2012**, *16*, 218–226. [CrossRef]
30. Ribeiro, J.S.; Ferreira, M.M.C.; Salva, T.J.G. Chemometric models for the quantitative descriptive sensory analysis of Arabica coffee beverages using near infrared spectroscopy. *Talanta* **2011**, *83*, 1352–1358. [CrossRef] [PubMed]
31. Ky, C.-L.; Louarn, J.; Dussert, S.; Guyot, B.; Hamon, S.; Noirot, M. Caffeine, trigonelline, chlorogenic acids and sucrose diversity in wild *Coffea arabica* L. and *C. canephora* P. accessions. *Food Chem.* **2001**, *75*, 223–230. [CrossRef]
32. Martín, M.J.; Pablos, F.; González, A.G.; Valdenebro, M.S.; León-Camacho, M. Fatty acid profiles as discriminant parameters for coffee varieties differentiation. *Talanta* **2001**, *54*, 291–297. [CrossRef]
33. Pan, L.; Lu, R.; Zhu, Q.; McGrath, J.M.; Tu, K. Measurement of moisture, soluble solids, sucrose content and mechanical properties in sugar beet using portable visible and near-infrared spectroscopy. *Postharvest Biol. Technol.* **2015**, *102*, 42–50. [CrossRef]
34. Williams, P. Grains and Seeds. In *Near-Infrared Spectroscopy in Food Science and Technology*; Ozaki, Y., McClure, W.F., Christy, A.A., Eds.; Wiley-Interscience: Hoboken, NJ, USA, 2007; pp. 165–217.

MDPI

St. Alban-Anlage 66

4052 Basel

Switzerland

Tel. +41 61 683 77 34

Fax +41 61 302 89 18

www.mdpi.com

Foods Editorial Office

E-mail: foods@mdpi.com

www.mdpi.com/journal/foods